The Science of Morality

Collected papers

Edited by
Graham Walker

Royal College of Physicians of London
2007

The Royal College of Physicians of London

The Royal College of Physicians plays a leading role in the delivery of high-quality patient care by setting standards of medical practice and promoting clinical excellence. We provide physicians in the UK and overseas with education, training and support throughout their careers. As an independent body representing over 20,000 Fellows and Members worldwide, we advise and work with government, the public, patients and other professions to improve health and healthcare.

Front cover
The cover features a design by Peter Maguire, © Peter Maguire.

Royal College of Physicians of London
11 St Andrew's Place, London NW1 4LE

Registered Charity No. 210508

Copyright
All rights reserved. No part of this publication may be reproduced in any form (including photocopying or storing it in any medium by electronic means and whether or not transiently or incidentally to some other use of this publication) without the written permission of the copyright owner. Applications for the copyright owner's permission to reproduce any part of this publication should be addressed to the publisher.

Copyright © 2007 Royal College of Physicians of London

Chapters reflect the opinions of the authors and should not be taken to represent the views or policy of the Royal College of Physicians.

ISBN 978-1-86016-286-2

Design, cover and page layout by Merriton Sharp, London
Typeset by Danset Graphics, Telford
Printed in Great Britain by The Lavenham Press Ltd, Sudbury, Suffolk

Contents

Contributors v

Preface vii

Foreword ix
AC Grayling

Introduction xiii
Graham Walker

PART 1: THE SCIENCE

1 **The neuroscience of morality** 3
  Baroness Susan Greenfield

2 **The neurology of consciousness – and conscience** 15
  Adam Zeman

3 **The evolutionary genetics of morality** 33
  Ian Craig and Caroline Loat

4 **The deceiving brain** 51
  Sean A Spence

5 **Values-based medicine: delusion and religious experience as a case study in the limits of medical-scientific reduction** 59
  Bill (KWM) Fulford

6 **Morality and the social brain** 81
  Robin Dunbar

PART 2: THE SOCIAL SCIENCE

7 **Demons and angels: who cares? Poor care structures and moral breakdown** 95
  Camila Batmanghelidjh

8 **Biological, psychological and social factors in the pathogenesis of psychopathy** 105
  Michael L Penn, Amer Pharaon and Atilla Cidam

9 **Universal values** 117
  William Hatcher

10 **The meaning of the 21st century** 125
   James Martin

Contributors

Camila Batmanghelidjh
Director, Kids Company, London

Atilla Cidam
Hackman Research Scholar, Franklin & Marshall College, Lancaster, Pennsylvania, USA

Ian Craig BSc MA PhD
Head of Molecular Genetics, Social, Genetic and Developmental Psychiatry Centre, King's College, London

Robin Dunbar BA PhD FRAI FBA
Professor of Evolutionary Psychology, School of Biological Sciences, University of Liverpool

Bill (KWM) Fulford DPhil FRCP FRCPsych
Professor of Philosophy and Mental Health, University of Warwick; Honorary Consultant Psychiatrist, University of Oxford; Co-Director, Institute for Philosophy, Diversity and Mental Health, University of Central Lancashire; Special Adviser for Values-Based Practice, Department of Health, London

Anthony Grayling MA DPhil(Oxon) FRSA
Professor of Philosophy, Birkbeck College, University of London; Supernumary Fellow, St Anne's College, University of Oxford

Baroness Susan Greenfield CBE MA DPhil DSc(Hon) FRCP(Hon)
Professor of Pharmacology, University of Oxford; Director of the Royal Institution of Great Britain

William Hatcher[†]
The late Professor William S Hatcher (1935–2005) held professorial chairs in mathematics and philosophy in Europe, North America and Russia

Caroline Loat BA(Oxon) MSc(Lon)
Postgraduate Research Student, Social, Genetic and Developmental Psychiatry Centre, King's College, London

James Martin MA(Oxon) DLitt(Oxon) DSc
Founder, James Martin 21st Century School, Oxford University; Founder and Chairman Emeritus, Headstrong, Inc, Washington

---

[†]Sadly, William Hatcher died before this book was published.

Contributors

Michael L Penn PhD
Associate Professor of Psychology, Franklin & Marshall College, Lancaster, Pennsylvania, USA

Amer Pharaon
Hackman Research Scholar, Franklin & Marshall College, Lancaster, Pennsylvania, USA

Sean A Spence MBBS BSc MD FRCPsych
Professor of General Adult Psychiatry, Academic Clinical Psychiatry, School of Medicine and Biomedical Sciences, University of Sheffield

Graham Walker BDS FDSRCS MB BS MA(Ethics)
Consultant Maxillo-facial Surgeon

Adam Zeman BM(Oxon) MRCP(UK) DM FRCP
Professor of Cognitive and Behavioural Neurology, Peninsula Medical School, Mardon Centre, Exeter

Preface

Science and morality may appear to some as incompatible yet they are entirely consonant. Science, in the words of Karl Popper,[1] is the search for truth, and some aspects of morality can be illuminated by science in the same way that science contributes to the understanding of, for example, the visual arts. The science of morality is a new discipline that addresses this dilemma and has applications in all spheres of human reality.

Our view of morality may derive from a wide variety of sources such as religion, natural law, mystical experience or philosophy. Moral codes are also time-, gender- and culture-dependent, so inevitably there are inconsistencies and contradictions between moral codes. This collection of papers explores some of these spheres. The first part is made up of science papers, and the second is made up of chapters from the disciplines of sociology, psychiatry and philosophy.

These papers are derived from a conference held at the Royal College of Physicians of London in 2002. 'The Science of Morality' conference was designed to bring together researchers from different disciplines whose brief was to consider the question of a scientific basis for morality and so provide common ground for diverse moral opinions. For some of the speakers, the subject was familiar territory, and for others it was completely new. It was an enlightening two days of sharing new knowledge within a constructive atmosphere. I am indebted to the speakers who found time within their busy schedules to prepare and deliver the papers collected here. I am also grateful for the invaluable help in producing this book given by Dr Peter Watkins, Ms Diana Beavan and the Royal College of Physicians Publishing Department.

GRAHAM WALKER
Editor

---

[1] Popper K. *In search of a better world: lectures and essays from thirty years.* London: Routledge, 1992.

Foreword

In the rapid and admirable growth of knowledge in biology, neurology, psychology, sociology and ethology there is rich promise, already fulfilled in a number of directions, for advances in understanding what Alexander Pope called 'the proper study of mankind', namely: mankind itself. One crucial area on which the various advances in these sciences might throw light is morality. At its simplest, the possibility is that by discerning the degrees to which human moral sentiments and practices are (as we say) hard-wired or soft-wired,[1] we will not only gain much understanding, but a better purchase on how to organise ourselves as a species, and therefore deal with the all-too persistent problems and deficits that bedevil us.

The discussions in this book address different aspects of this vital possibility. The following observations offer a background to them, touching on the reasons why it sometimes seems that most light for understanding the basis of morality comes from the hard-wire direction, while at other times it seems to come from the soft-wire direction. And this of course poses a challenge to investigate how it might after all come from both in interaction.

Morality is a matter of relationships. It concerns the obligations and responsibilities that people have towards one another individually and to their community at large. It also concerns obligations people have to themselves, and – as we have come to realise very recently in the Western tradition of thought – to other animals and the natural environment in general.

The questions on which moral reflection focuses concern the right way to live and behave. How should one treat others, and respond best to their needs and interests in the light of one's own needs and interests? What things matter – what things are truly valuable, and conduce to the good – in our choosing and acting? What are our duties to ourselves, and how do we justify (if we can) treating those closest to us differently from how we treat strangers? What anyway is 'the good' and how does one judge what is right in circumstances of dilemma? Such questions are the staple of moral philosophy, and answers to them are the staple of systems of morality which dictate or at least enjoin particular sets of attitudes and practices.

People regulate their interactions not only on the basis of what they recognise as moral considerations, but also – more weakly – by traditions of etiquette and social customs, and – more strongly – by enforceable legal systems with

---

[1] That is, instinctual or learned.

punitive sanctions attached. Of course, sanctions for breaches of moral codes can be very powerful too, involving social rejection and opprobrium; and even in the case of customs and etiquette there can be sanctions, as when people avoid the boor and the bore. The three modes of mores, morals and law overlap and sometimes entangle, but generally speaking it is not hard to distinguish what counts as a matter of morality for a given society, nor is it always hard to recognise when aspects of a moral outlook have been turned into positive law (as for example in societies where adultery is legislated into a criminal offence).

Historically one can see how value systems reflected the identifiable needs of a given society at a given time. In pre-classical times virtue (where the first syllable of this Latin-derived term, 'vir', specifically denoted the masculine) attached to men of courage, fortitude, endurance and martial skill. Hector and Achilles were the admired models. In the classical age, as the teachings of Socrates and the dramas of Aeschylus between them show, there was a turn away from these warrior virtues to civic virtues of probity, cooperation and continence. The warrior virtues remained necessary for times of danger, but were disruptive in settled conditions of peace, and with the growth of urban civilisation needed to be turned from destructive into more constructive channels. One can follow this transition in the literature and philosophy of the time; they are foundational for Western culture in general.

Likewise the origins of sexual morality in the Judaeo-Christian tradition can be traced to the exigencies of the herdsman's condition. The life and death importance of increasing one's flocks made reproductive considerations paramount. Any misdirection of seed – as with Onan, or in homosexuality – was a crime against the chances of surviving, let alone flourishing. But the practice of concubinage, of fathering children on slaves or maidservants, and even adultery as in the story of David and Bathsheba, was widely countenanced in Old Testament morality, because they promised reproductive increase.

All these thoughts push one towards thinking that morality is a social and historical soft-wired matter, acquired through socialisation of the young into the attitudes and practices of the communities into which they happen to be born. The differences in moral outlooks as between different cultures, and within different historical phases of the same culture, add further weight to this view. How can it be that (say) homosexuality was condoned in ancient Greece, condemned in Christian morality, and accepted again in modern secular polities, unless its place in the moral economy of a society is a matter of choice and discussion?

On the other hand, human beings are essentially social animals – where 'essentially' has full force as meaning that the social nature of humanity is one of its defining characteristics. And this is a consequential fact. The social instincts are powerful, and fully present in the earliest stages of life. A neonate's ability to mimic the facial expressions of a carer, any normal person's ability to recognise and respond instantly and in fine-grained ways to the emotional states of others, the profound need that individuals have for intimacy and companionship, the deep sexual impulses and the obviously chemical basis of infatuation and romantic love – despite poetic objection to the thought that this latter is connected to the synthesis of an amino acid peptide in hypothalamic neurons, and conveyed into the lover's hot blood down the axons of the posterior pituitary – together with a number of other such findings, conspire to suggest that the genetically-determined architecture of the central nervous system is the habitation of a great deal of our moral capacities and responses.

The findings of ethology, and especially studies of primate behaviour, reinforce this thought. Nurture is not absent from the formation of chimpanzee behaviour, but in comparison to human beings where large cultural differences can only be ascribed to nurture, nature is the obvious trump-holder among the other apes as it is in the rest of the animate world. Work in sociobiology and evolutionary psychology, for all that it is controversial, adduces much compelling evidence to show that considerations about how human beings interact and respond to each other, and how they build the fabric of their societies as a way of managing these interactions, has roots that appear to run deep into the biological past of the species.

Moreover the lessons learned from studies of deprived and abused children show that deficits in the meeting of their hard-wired needs typically leads to distortion in subsequent emotional life, and adult inability to relate to others in well-adapted and socially fruitful ways. Here the influence of nurture on nature is unhappily apparent, but with the emphasis placed on the fact that nature is the material on which nurture operates, and therefore sets the parameters for what nurture can produce. At the same time, studies of the effects of injury and disease on the brain show that much of what moral interaction rests upon – for example, such basic things as facial recognition, processing of speech, ability to empathise – is wholly contingent on the integrity of the brain's functioning, and accordingly owe themselves to the existence of the relevant structures in the first place, these having evolved through biological time as features of our genetic endowment.

Against these thoughts is the competing observation that a great deal of morality is aimed at the control, even the suppression, of natural urges and inclinations evidently lodged in this hard-wiring. Anger, aggression and sexual desire are examples of nature at work in human beings, and the management and channelling (in some theories: sublimation) of the impulses involved are the work of conscious choices, much education, and cultural direction and correction. Here the influence of intelligence – mankind's most distinctive and influential evolutionary adaptation – and moreover collective intelligence, which is the result of historical experience and consciously directed debate, is manifestly at work, helping to reload the dice on the nurture side of the story.

The discussions in the pages to follow canvass the whole range of these considerations, and more. They move from the brain and its extraordinary product of consciousness to psychological and sociological insights, and they examine from the breadth of these perspectives the possibilities for increased understanding of the social and moral capacities of human beings. There can scarcely be a more important set of enquiries. Above the portal of the oracle at Delphi in ancient times stood the inscription, 'Know Thyself', and this book suggests that we are beginning to do just that, especially if the right relationship between nature and nurture, genetics and culture, brain and the thought it secretes, can be determined. It has to be said, though, that as the final discussion here shows in relation to the fragile state of the world caused by various human immoralities, this progress in responding to the Delphic injunction is happening not a moment too soon.

AC GRAYLING
Professor of Philosophy, Birkbeck College, University of London
February 2007

Introduction

GRAHAM WALKER

Morals are the principles which guide personal behaviour. They equate with ethical theory which should ideally be based on universal values. Moral codes may be derived from any combination of natural law, philosophy, altruism, utilitarianism, mysticism, tradition and theology, so it is unsurprising that currently there is no universal moral system. With no objective foundation, it is easy to disguise prejudice and intolerance with what is offered as defensible moral judgements or to paralyse discussion and questionning by phrases such as, 'It is in the book'. This moral relativity is a frequent source of conflict.

The papers in this book concern the scientific foundation of our sense of morality and about how that moral relativity could be reduced. For example, 2,000 years ago the population of the earth was 200 million and the doubling time was 1,000 years. It was moral to go forth and multiply, notwithstanding some of the social implications. Today the population is around 6.5 billion and the doubling time is 50 years. Few would consider the same exhortation to be moral now.

Morals are also gender dependent. We are aware of the different male and female roles that society has shaped, only some of which are for supportable reasons. But these different roles should not confer different civil or human rights. Honour killings and stoning female but not male adulterers to death are reasoned disincentives to protect morality in some cultures, while in others adultery is accepted if not condoned. 'Universal morals' apply to both genders equally.

Cultural moral relativity causes more problems as cities become multiracial. Imbibing alcohol is seen as immoral hedonism by one, but is a harmless pleasure, almost a rite of passage, to another.

Morality can also be numerically dependent. As Steven Pinker states, recalling a personal communication from Donald Symons in *The blank slate*:

> If only one person in the world held down a terrified, screaming little girl, cut off her genitals with a septic blade and sewed her up leaving a small hole for urine and menstrual flow, the only question would be how severely should that person be punished. But when thousands

commit the same procedure on millions of girls, the enormity of the act is not magnified a million times. It is instead, attributed to culture and magically becomes less, not more, horrible.[1]

The field of morality is of shifting sand and its principles are sometimes artfully applied. Moralising of political or ethnic matters has frequently been utilised to license aggression against those with whom we disagree. Interpretation of holy scriptures by the learned for the purpose of manipulation of the public is commonplace. It is therefore highly desirable that our moral uncertainty should be relieved and moral cohesion established.

How does science affect moral relativity?

Science can not only help in dispelling suspicion, misinterpretation and harmful dogma, but can also provide some evidential basis of a non-denominational, consistent morality founded on universal values. It is not a new ontology – 'macro' aspects of this relationship between science and morality surround us and are obvious and historic. Promiscuity is related to divorce, illegitimacy and sexually transmitted disease; greed and intolerance are related to violence and war; such social evils are clearly related to absence of moral values. What is new and less self-evident is the 'micro' aspect and this information has only become possible because of the enhanced understanding of genetics, neurology and physiology afforded by technological advances. For example, science is able to explain the microbiological cause of sexually transmitted disease and therefore the relation to promiscuity which moral teaching forbade. The intellectual disintegration associated with drug abuse is explicable by the neurological degradation diagnosed on brain scans or biopsy. The effect of the media on child behaviour is clear only when the large numbers are analysed by computer. Science makes the case for morality more believable. However, science must also be seen as a relative truth. Some of what we positively believe as undeniable truth today will be discarded in the next 10 years, in the same way as we have discarded some scientific 'certainties' of yesteryear. Science and morality both share relativity but whereas moral relativity is the cause of disagreement and even extreme violence, scientific relativity is accepted and statistically incorporated into thresholds of certainty.

Hard and soft science

The appeal of a scientific basis to morality is multifaceted. Primarily, science is replicable and objective. Scientific methodology is therefore a cohesive force that unites scientists, and a respected source of knowledge for the public. More importantly, it supports a common moral perspective for every culture, religion and race by helping to dispel superstition and fear. This book examines some aspects of the neurology of morality provoked by certain questions. Is there a brain centre for morality? As Baroness Greenfield (Chapter 1) shows, loss of consciousness during anaesthesia is not associated with inactivity in any particular area: all areas of activity close down. As she explains, moral deliberation relies on conscious thought and we can therefore conclude that morality is not related to a centre but to the whole neuronal network.

How does hard neuroscience sit with sociological and behavioural science? The late William Hatcher (Chapter 9) points out that the simple fact that all individuals react positively to love, acceptance and generosity proves the universality of what he calls spiritual values. It may be argued conversely that some races exhibit different emotional reactions and demonstration of these three behaviours in certain circumstances may be entirely unwelcome. However, if we define emotional reaction in terms of the intellectual content, not the stimulus, all races and cultures will exhibit literally hundreds of common behaviours. So, if a person feels pleased, then the facial expression would be identified by other races and cultures as indicating pleasure. The stimulus will vary with culture, however; I might applaud a fine golf shot and show pleasure but to an Amazonian Indian the same experience would be greeted with puzzlement.

Brain development and morality

Electron microscopy of neonatal brain tissue shows few neurones with few synapses on a featureless matrix. At two months old this has totally changed. There is now a rich network of neurones and a myriad of connections. At two years old the picture is almost all neurones and synapses. This change is the anatomical result of learning: more facts lead to more synapses. The network allows numerous connections to be made, some repeated reflexively, as in riding a bicycle, some new as in innovatory thought. This is the all-important capacity known as neural plasticity. It is reasonable to expect those networks to reflect the type of experience and learning. If a child is subjected to violence and deprivation, there will be synapses registering this experience. One surrounded by love and comfort will have formed different synapse groupings. These groups form the basis of associated memory and can be elicited by psychiatrists with word or picture association tests. To one patient, red is associated with rose, love, happiness. A different upbringing may retrieve the sequence of blood, pain, hate.

The gradual elaboration of memory and reasoning proceeds at a particular pace. Psychometric testing shows that levels of sophistication of reasoning are age related. An infant would not be aware of much more than sensations: heat, cold, hunger and so on. A five-year-old would be expected to be protective towards a sibling but not to appreciate why excessive consumption is destroying the planet. Six stages of childhood development were described by Jean Piaget, the Swiss psychiatrist. This developmental process can also be described in terms of levels or orders of thought. For example, 'I know the date' is first order of thought. 'I know that you know the date', is second order of thought, and so on. Higher primates are capable of second order of thought, as is a five-year-old human. This capacity is one of the prerequisites of moral deliberation. We do see this in the apes and orang-utans. Robin Dunbar indicates that second order thought enables apes to hold details of acceptable behaviour, who to please and who it is safe to bully, who owes or is owed a favour: a basic moral code in fact. The communal discipline endows the group with stability which in turn allows peaceful aggregation of larger numbers. In contrast, the larger brain capacity of humans allows fourth order thought and therefore the intellectual equipment to moralise beyond self to a moral system which embraces all humanity, detached altruism, the future of the planet and so on.

Sophisticated levels of thought, however, can be employed for good and ill. Sean Spence (Chapter 4) studied deception in humans by observing the difference in response delay to a question answered truthfully or dishonestly. Volunteers were instructed to answer questions truthfully or otherwise while the brain function was monitored on a magnetic resonance imaging (MRI) scanner. This showed a delay in the lying situation and thus that the default state of the brain was truth telling. This is not surprising since truthful response is first order thought and to deceive requires second order thought and a period of deliberation. Lying is therefore an acquired social skill but what purpose does deception serve? Good and bad would be the answer. To always tell the truth would be difficult and occasionally brutal, while lying occasionally smoothes social intercourse. On the other hand, deception may be used for self gain. To deceive for the purpose of protecting a friend's sensitivities is acceptable but deception for gain causes offence. For the deceiver, it may confer only short-term advantages, as we see from studies of game theory.

Game theory first used single situations to test altruistic or selfish responses but it was then theorised that life was not a sequence of single games but similar situations which were repeated. This introduced the elements of trust, forgiveness and reputation. Then the outcomes were more optimistic, for the altruists prospered if the game was played long enough. The extrapolation of game theory to human behaviour must be limited but it confirmed the social value of a reputation for honesty, altruism and justice. It also indicated that reciprocity and punishment were useful social tools. Game theory indicates how these factors are advantageous and some of the scientific data show that they are predisposed, either by nature or nurture.

In survival games of various designs, where players are secretly given roles of cooperators, 'honest johns' or liars, the honest cooperators usually win through and achieve a stable community. In primates and primitive humans this community confers advantages of protection, cooperation, larger groups and a richer gene pool to reduce harmful mutations. The intellectual capacity to enable this can be related to the size of the neo-cortex, also named the social brain by Sean Spence. More specifically, the social brain resides essentially in the orbito-frontal cortex.

Moral impairment

Morality thus clearly depends on brain capacity as well as experience and it may be impaired by damage to the social brain. Such damage may occur through physical injury, tumour growth, degenerative disease, for example Alzheimer's disease, and substance abuse. The first recorded case of behaviour change after social brain trauma was reported in 1856 when an American railroad worker by the name of Phinneas Gage was compacting dynamite into the side of a hill with an iron bar prior to excavation. The dynamite exploded propelling the metre long iron bar through the left cheek, eye socket and frontal part of the brain. After a brief period of unconsciousness, Gage sat up and was helped to the local hotel. He recovered sufficiently by six weeks to return to work. However, his doctor recorded his change in personality from a gentle, sociable and good husband to a violent, addicted, dishonest man.

The frontal cortex receives information from, among others, the hard-wired centres, from memories and directly from the centres that register sensory experiences. It probably processes this input, deliberates and formulates patterns of behaviour and judgements. Such formulations are normally subject to updates as new experiences or thoughts are logged. The brain retains this malleability, the capacity to independently evolve thought processes following new information or meditation. In highly artificial circumstances such as indoctrination, this independence of thought can be disabled and certain thoughts become inaccessible to reason and logic. This is the mental process of fanaticism and is typically used to promote the importance of dogma above the value of the individual. In the Kantian concepts of ethics, this is the elevation of means above the ends. An example is the death of millions during the establishment of communism.

Let us take the question of child training a little further. Why do apparently healthy children sometimes become sociopathic? Michael Penn and colleagues (Chapter 8) argue that unless a child is able to recognise that there is a relation between misdemeanour and punishment, that is to have a sense of consequence, they will be unable to respond to the rationale of discipline. This may arise in two ways. It may arise if the child has a genetic trait which predisposes to antisocial behaviour or it may arise by repetitive irrational abuse. In laboratory simulation of the former, a group of normal adults and psychopathic adults were informed that at the end of a 10-second countdown they would receive an unpleasant stimulus. Skin electro-conductivity measurements in psychopaths showed little change in conductivity while non-psychopathic volunteers showed a significant rise in conductivity when the countdown started and an increasing rise as the count approached the end-point. The physiological change associated with fear of anticipated pain is absent in psychopaths. The neurological defect seems to be in the limbic system of the brain, a centre which is strongly associated with emotional development.

Another experiment (also described in Chaper 8) that added to the understanding of training involved the state of helplessness. Two groups of volunteers were subjected to aversive events such as painfully loud noise. One of these groups, by application and perseverance, could find a method of ending the aversive event and applied themselves to solving the problem of each event. For the second group, there was no solution and they eventually realised that they could exert no control and began to suffer the aversion passively. The third group were subjected to neither and were simply the control group. It is suggested that the real-life equivalent is the recognition of relation of action to outcomes. The important outcomes which shape moral behaviour are reward and punishment. If there is a neurological deficit which prevents this logical modification of behaviour, antisocial behaviour disorder is the likely outcome.

When there appears to be no such justice or logic to life's successes or failures, where good behaviour or bad have the same result, then again the outcome is likely to be antisocial behaviour, as shown by Camila Batmanghelidj (Chapter 7). Management of the two aetiological types, however, is quite different. Here we have an interesting philosophical point relating to forensic factors. The definition of the antisocial behaviour disorder is independent of causes. If the diagnosis relies on definitions and protocols that are evidence based then there may be no difference in the management of the offender. If there is a value-based diagnosis there should be consonance between treatment of antisocial behaviour and cause. There may be a group with neurological deficit due to trauma or genetic abnormality which cannot be culpable, since they are not conscious of the digression and incapable of responding to corrective discipline. The group who are antisocial because of abuse are capable of response but need expert therapy.

The media is often held responsible for engendering violence in children and for its continuing expression in adulthood. There are several mechanisms suggested, the most supportable of which are desensitisation (repeated viewing of violent material), behavioural effects (encouragement to see aggressive behaviour as the norm), and culturation effects (developing a distorted view of the world). While there is a significant body of opinion that supports the relationship between violent behaviour and the hours of violence viewed, the Commission on Children and Violence (1995) found that the context in which the violence was viewed was the primary factor. This indicated that a child in a moral domestic environment would be less influenced than one in a permissive or violent environment.

Genes and morals

There is a variable relationship between genes and behaviour abnormalities. Schizophrenia has, on the one hand, long been recognised as an inherited abnormality. For example, if one identical twin is schizophrenic then the relative risk for the other will be 50%. If one fraternal twin has schizophrenia the other has a 25% chance of being affected. However, according to Sawa and Kamiya,[2] this disease is a neuro-developmental fault involving architectural, cellular biological and protein abnormalities, which could all be due to a gene for schizophrenia. But there appear to be additional influences, which are involved in the full manifestation of the symptoms of schizophrenia.

So, even for schizophrenia, there appears to be a multi-factorial influence on expression which demotes the importance of the gene. Hence the behavioural geneticists' assertion that most genes are probabilistic. That frustrates the nature versus nurture antagonists, for neither can claim pure expression of effect, not even for schizophrenia. The controversy between nature and nurture is not so swiftly resolved by the assertion that there are more subtle and less predictable effects due to combinations of genes on adjacent sites and variable penetrance. The expression of some genes may also depend on whether it is inherited from the father or the mother. To complicate the picture further, the expression of any gene, not just schizophrenia, may be influenced by the environment. As Dr Venter, President of Celera, the American company which defined the human genome simultaneously with the Sanger Laboratory in Cambridge, said 'You cannot define the effect of genes without defining the effect of the environment'.

Clearly, the determinist view of one gene per character for humans cannot be correct with merely 30,000 genes, only twice as many as the fruit fly, even if the environment does influence expression. Equally, the reductionist view that understanding the genome will allow complete explanation of human variability is over-optimistic. How then, can the sophistication and variation of the human being be explained? The answer is in the variable penetrance, environment effect, group effect, subtle control genes and a large numbers of mini-genes or snips (single nucleotide polymorphisms), which confer multi-functionality. The frustrating thing for eugenicists is that the qualities of giftedness such as great musicianship, athleticism or leadership are emergenic, expressed only when there is a certain combination of genes and circumstantial factors.

This view seems obviously true when one confronts the myriad complexities of the effects of the FOXP2 gene for example,[3] which plays an important role in language and speech development and is one of a family of genes involved in the formation of the embryo itself. Its influence is therefore exerted at both ends of the spectrum of development, embryological and social. In conferring the capacity for speech and therefore the ability to share knowledge and experience, to organise and cooperate, it was the gene which probably enabled socialisation more than any other, with a fundamental role in evolution of moral behaviour.

Therefore, there are some genes which have a narrow predictable expression and others which are more variable and subtle. Ian Craig and Caroline Loat (Chapter 3) explain that one example of the former is the DRD4 gene mutation, the effect of which is antisocial behaviour and aggression. It seems to have arisen about 40,000 years ago and was probably advantageous to survival, then spreading quickly through the population; unsurprisingly, its expression is sex linked.

The description of the human genome and thus the fundament of humanness can be written down as a formula which is common to all. This in itself has underpinned the oneness of mankind and the elementary observation that the only qualification for human rights is to be human, not to be a particular colour or race. Mining the human genome treasures will augment the objective understanding of human behaviour.

Some single genes have a powerful influence on mind and have a dominant expression. Most traits, however, are related to groups of less dominant genes more or less influenced by other genes. There are genes related to moral behaviour but of subtle expression and therefore easily overwritten by upbringing. Many studies have on the other hand shown that the behaviour of profoundly antisocial individuals has shown consistency since early childhood through adulthood and defeated the best efforts of their parents. This again suggests an inherited causation.

The future

We have seen how science may explain our innate moral nature and how it might be evolved or compromised by experience, genes and brain damage. This collection of papers is a sample of the huge volume of available science which can be adduced to this topic. Now a word about the future. Carbon-silicone interface chemistry promises implantable data and intelligence. Silicone chips loaded with information and programs may be implanted in humans, short circuiting the onerous task of learning. Buy a chip off the shelf for a law degree, an international language. However, there are many reasons for caution. Who loads and programs the chips? It is unlikely that any ethical body would be funded for the costs of development even if such advances were acceptable. This kind of project is entrepreneurial and that leaves open the door to subliminal suggestion for political or commercial reasons and loss of autonomy. Another point is that data implantation does not endow wisdom. We might argue that eliminating the hours of learning information which is readily available from implants, or, less controversially, powerful computers the size of a wrist watch, would free us to use the time more effectively in the deliberation which begets wisdom.

References

1. Pinker S. *The blank slate*. London: Penguin Books, 2002:273.
2. Sawa A, Kamiya A. Elucidating the pathogenesis of schizophrenia. *BMJ* 2003;327:632–3.
3. Lai, CS, Fisher, SE, Hurst, JA, Vargha-Khadem, F, Monaco AP, A novel forkhead domain gene is mutated in a severe speech and language disorder. *Nature* 2001;413:513–23.

Part 1
THE SCIENCE

1

The neuroscience of morality

BARONESS SUSAN GREENFIELD

Professor Greenfield refers to a wide spectrum of neurobiological research in her dismissal of the notion of a brain centre for the conscious. The relevance of input of information, synaptic proliferation and deliberation to the evolution of a moral awareness is explained. This chapter touches on the effect of ageing and drug abuse on cerebral competence. Future challenges to autonomy and ethics arising from technological developments such as nanotechnology are also provocatively discussed.

The neuroscience of morality might be seen as a contradiction in terms, as some see the brain, or people who work with the brain, as being amoral anyway and therefore irrelevant to the science of morality. However, I shall argue that science and morality can be linked in two ways.

First I shall explore the nuts and bolts of the brain and see how these may accommodate morality. Then, perhaps more chillingly, I shall look to the future and explore what impact the new technologies will have on our brains and minds, and how we might respond from a moral perspective in a future society.

What is needed for a protective morality? Two basic requirements appear to be a sense of self and a sense of consequence of one's actions. Another question worth asking is: why do our brains and minds differ from those of animals? This is an important issue because all too frequently the 'mind' is conflated with another term, 'consciousness'; so I shall attempt, as a neuroscientist, to distinguish between those two terms.

What and where is consciousness?

Let us start by locating consciousness. If we are really adopting a physical approach to this, then clearly it must be somewhere in the brain. Is there a centre for consciousness? We need to define it, but instead of using the whole chapter to describe and define consciousness, I will say immediately that it is your own personal world that only you can access; it is the thing you lose when you sleep. But is there a centre for consciousness of, say, relations and another for emotions? Is there a mini centre for love and another for hate within the emotion centre? The problem with this infinite regression is that we miniaturise the problem but do not solve it by positing that there are brains within brains and that the brain is compartmentalised into autonomous brain regions. No one believes this but even neuroscientists will sometimes lapse into listing centres for this and centres for that. I suggest that the concept of a centre for consciousness makes no logical sense and does not help to solve anything. Figure 1.1 shows brain scans of patients' brain function during anaesthesia. While the five subjects are awake there are multiple areas that are working, and following halothane anaesthesia there is a uniform shutting down of all regions. There is no single area that is extinguished.

However, if there are non-committed brain regions which are somehow working together to give us consciousness, they do not create multiple consciousness. Figure 1.2 is a very familiar picture showing an old crone from one perspective and an image of a 19th century young girl from another. The point of this is that you can either see one or the other but you cannot see both at the same time. So consciousness is spatially multiple but in temporal terms it has unity. So we have to look for something in the brain that allows non-specialised neurones from all regions to work together to give a temporal state of unity.

Another issue which has an ethical association and moral context is whether a fetus is conscious. We know that the manner of birth can vary; you can either

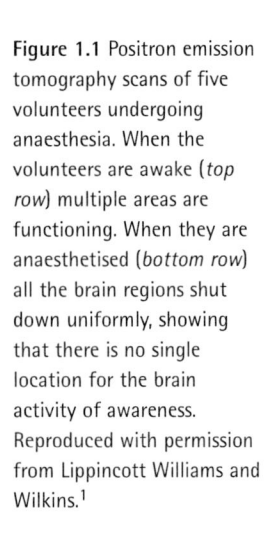

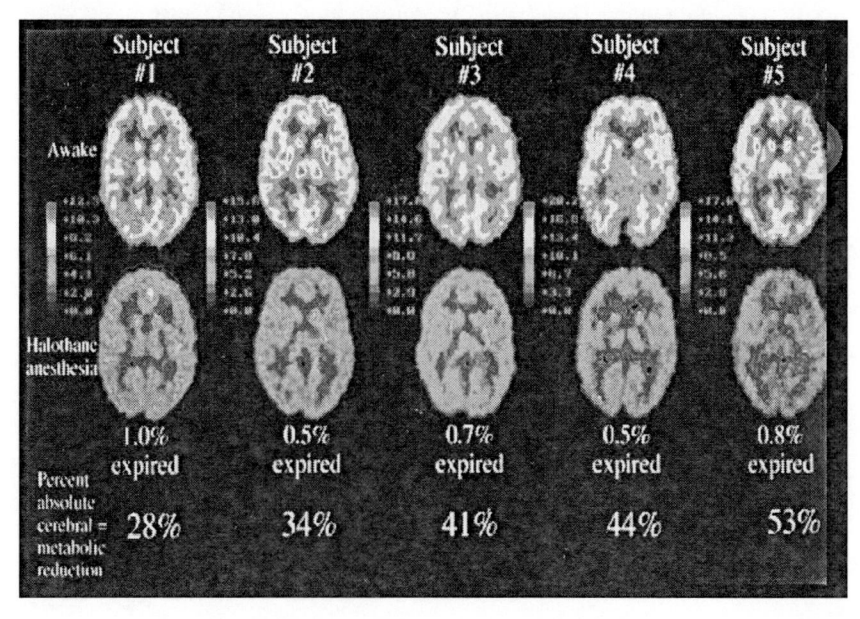

Figure 1.1 Positron emission tomography scans of five volunteers undergoing anaesthesia. When the volunteers are awake (*top row*) multiple areas are functioning. When they are anaesthetised (*bottom row*) all the brain regions shut down uniformly, showing that there is no single location for the brain activity of awareness. Reproduced with permission from Lippincott Williams and Wilkins.[1]

Figure 1.2 The picture shows an old crone or a 19th century young girl, depending on the perspective of the person looking at the picture.

have a caesarean section or give birth via the birth canal. The timing can also vary: you do not have to be born exactly at nine months. So it is very hard to think of the manner of birth or the timing of birth as a Rubicon which you cross and suddenly your brain acquires consciousness. As we all know, the brain does not physically change suddenly at birth. So, does the baby become conscious some time after birth, and if so, when? At a month old or a week old? It becomes absurd trying to guess when the fetus becomes conscious, before or after birth. I would suggest, controversially, that the fetus is conscious, but also that consciousness is not all or nothing but grows as the brain grows. So, as the brain develops in a fetus, then consciousness also develops and as far as animals are concerned the more sophisticated the brain, the more conscious the animal. The rat is thus not as conscious as the dog, the dog is not as conscious as the primate, and the primate is not as conscious as the human being. Furthermore, you as an individual will be more conscious at some times than at others. We acknowledge different levels of consciousness when we talk about raising or deepening our consciousness.

So I have turned something ineffable, elusive and immeasurable into something that can be measured. Because if something develops by degree then we can revisit the brain and see what is getting bigger or smaller from one moment to the next. Finally, touching on the issue of intentionality, I suggest that we are always conscious of something. So, consciousness is spatially multiple yet effectively single at any one time. It is an emergent property of unspecialised groups of neurones that are continuously variable with respect to some kind of epicentre, some kind of trigger. As the group of neurones expands so does the depth of consciousness and hence the potential for self-consciousness. I suggest that self-consciousness is part of the continuum of consciousness that entails a very large degree of it. So imagine a stone being thrown in a puddle, which is the trigger that will start transient highly evanescent ripples, and the ripples could be proportional to the degree of your consciousness. If we can find a neural equivalent of a large stone versus a small stone, a forceful throw versus a gentle throw, viscous flow versus water, and competition from other stones, then we can use that model to tease out different variables that will determine the depth of your consciousness and hence your self-consciousness and your prevailing sense of morality at any one moment.

The most obvious candidate for a stone in the puddle is a neurone, of which of course there are billions. For example, if I see my mother, that could fire my 'mother-recognition' neurone, which triggers the ripples. Sadly, we must reject the concept of a mother cell because if I had never seen my mother it would be a waste of a cell, and conversely, if it died halfway through my life I would no longer know my mother. So I propose a theory midway between single-cell theory and the idea of large fixed-function brain regions: that is, the idea of neuronal networks. Figure 1.3 shows a picture that looks like a jungle which is a sensible metaphor to use where you have a single brain cell but up to 100,000 connections on to any one neurone. If they were counted at a rate of one per second it would take 32 million years to count them all. We know that these are highly dynamic connections, as shown by a five-day experiment with human subjects, none of whom could play the piano (Figure 1.4). They were divided into three groups:

- The control group just played around with the piano keys. Their brain scans over a five-day period show that little happened in the brain region related to digits.
- The next group were performing five-digit piano exercises. In five days of learning these exercises, their brain scans reflect that experience.
- The middle group learnt the five-finger piano exercises but only mentally: they were allowed only to imagine the exercise but not to perform it. However, their brain scans show significant changes.

This experiment disproves the misleading distinction between mind and brain and mental events which are often held to be unrelatable, and grounds them all in biological processes.

1 The neuroscience of morality

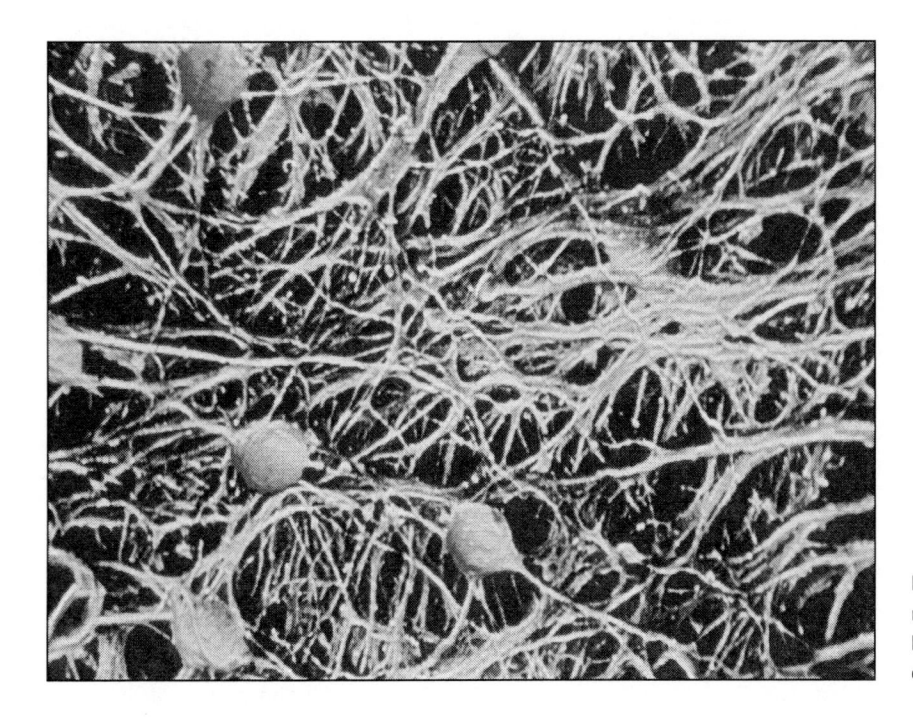

Figure 1.3 Neuronal networks. A single neurone has up to 100,000 connections.

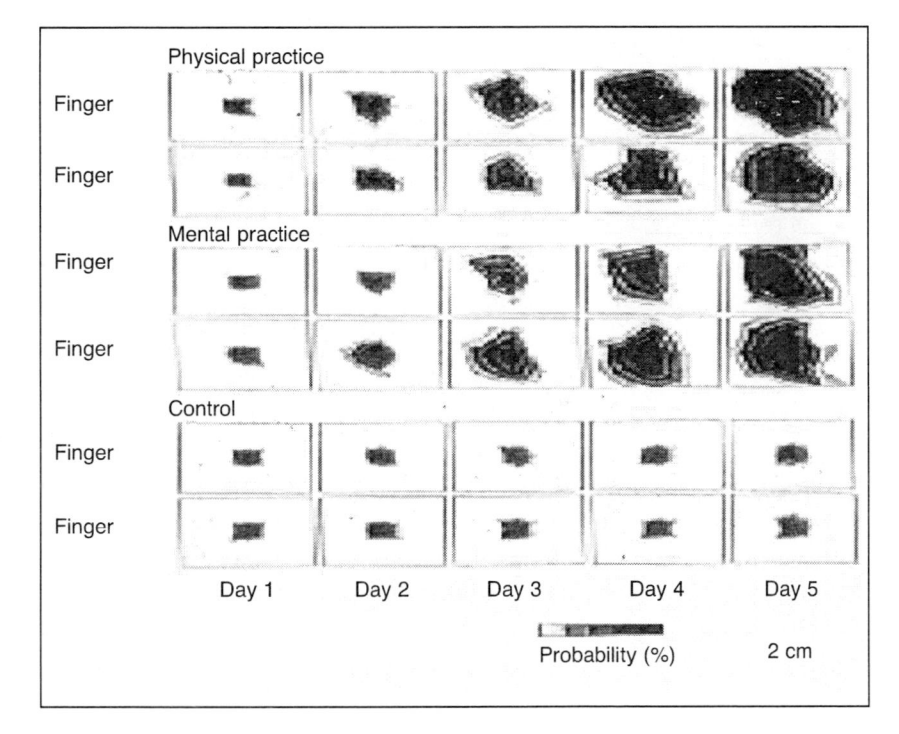

Figure 1.4 A five-day experiment showing the brain activity of three groups of people, none of whom could play the piano. *Top*: a group who were performing five-finger piano exercises. *Middle*: a group who learnt the five-finger exercises but were allowed only to imagine the exercise mentally, not to perform it physically. *Bottom*: a control group who were allowed to play around on the piano keys but were not given any piano exercises. Used with permission from the American Physiological Society.[2]

The hub of consciousness could be a network of brain cells which is hard-wired, experience based, highly local and quasi-permanent. How does it become activated to trigger a large assembly of brain cells? That leads to the question: what is a mind? We are born with almost all our brain cells, and the multiplication of these connections accounts for the growth of the brain after birth: they reflect our experiences and influence our further perceptions. We are born into a world where everything is evaluated in terms of how sweet, how fast or how bright, and gradually those abstract sensations coalesce into meaningful objects such as people. They will then feature in certain experiences and memories and therefore acquire a meaning according to the amount of associations that they trigger. This will remain highly dynamic so we go through life in a dialogue with the outside world, interpreting it in terms of what we have experienced.

The mind: a personalisation of the brain

So I suggest that the mind, far from being the vague alternative to the brain, is the personalisation of that brain. The growth of the brain after birth is due to the proliferation of connections rather than an increase in neurones. This proliferation reflects changes in self, personalising it so that even if you are a clone, ie an identical twin, you will have a unique configuration of brain cell connections. This is happening all the time: we are going through life with our personal hard-wired circuits and this leads to the ability to be self-conscious and exercise moral judgements. A very small child does not have that ability – it does not have those connections or these sophisticated processes. I would argue that our sense of morality is laid down and embedded in the proliferation of a unique configuration of connections reflecting our experiences.

We can lose our minds, as sadly happens in Alzheimer's disease, and senility can amount to atrophy of the dendrites, the branches which enable neurones to form those multitudes of connections. We can blow our minds with chemicals and remain conscious but have the minds of small children. In that instance, we have put ourselves in an environment that is evaluated in terms of sensory factors – the beat, for example, with no cognitive content whatsoever, where the music has no meaning, there are flashing lights and drugs like Ecstasy modulate brain functions. In circumstances like these, the brain can be manipulated to render some of the circuitry inaccessible. It is similar perhaps to road rage or *crime passionelle* or a scenario where, in the heat of the moment, you are a passive recipient of your senses. You are not accessing those connections and it is important to look at what the brain does in these different states in the adult, especially when we are considering morality.

The workings of the brain

Most of the time that our adult minds are working, our personalised world has a meaning because our circuitry is functioning. How can neuroscience contribute to an understanding of self-consciousness and mind and a sense of things that are perhaps not shared by creatures with simpler brains but that are nonetheless conscious? One pressing issue is the influence of future technologies on this mind, the most obvious one being genetic modification. Figure 1.5 shows DNA. In our excitement over identifying specific genes for criminality, homosexuality etc, we forget that genes are supposed to exist inside this substance that looks like cotton wool. In the same way that we cannot attribute a sophisticated function to a single region of the brain, I would posit that we cannot attribute specific characteristics to the lowest level, the gene. Let us see how the two match together. We have consciousness, an inner state, which we cannot really define. Of course if it goes wrong, if consciousness malfunctions, clinicians will be able to describe certain syndromes such as schizophrenia,

Figure 1.5 A purified sample of DNA which has been extracted from cells and stored in an alcohol solution.

which entails impairments of functions, be they sensory or cognitive. Those various functions can be broken down into sub-functions, for example in terms of colour-, form- or motion-processing, all of which are divided up in different brain regions, so that any one brain region will participate in many functions. Conversely, any one function will be distributed among many brain regions. Within each brain region we have circuits of neurones, the building block of which is the synapse. Across the synaptic gap, transmitters allow contact between one brain cell and another. That transmitter is synthesised, interacts with its own receptor and is degraded. All these agents and their enzymes are the product of gene-directed proteins. But you cannot make the leap from a gene to a final function. After all, we have 30,000 genes or more in our bodies, but there are about 1 billion more connections in our brains than genes in our chromosomes. There is just not enough genetic material to go round, so we have to look at these connections to see how else they can be influenced.

Let us consider Huntington's chorea, which is one of the very few examples of a single-gene disorder of the central nervous system. Figure 1.6 is a brain coronal section showing the loss of tissue adjacent to the ventricles characteristic of Huntington's chorea. In a single-gene disorder one might assume it would be possible to map a single gene to the disease. However, Van Dellen and colleagues[3] in Oxford showed that even then nurture could trump nature. Transgenic mice that have been engineered to have the murine equivalent of Huntington's chorea could nonetheless be protected to a certain extent by environmental stimulation. The onset of their disease was later in

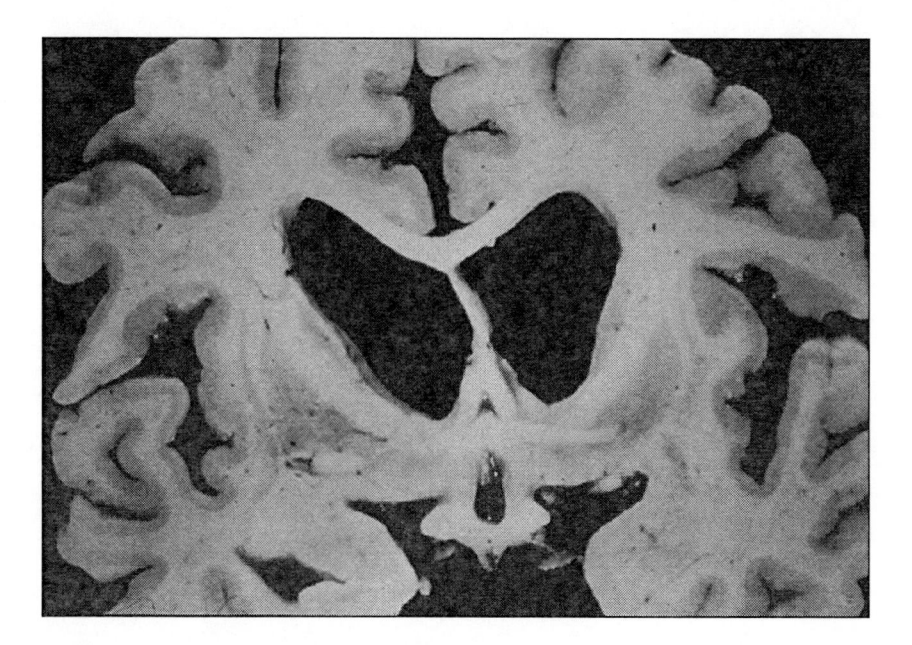

Figure 1.6 A brain coronal section showing the loss of tissue adjacent to the ventricles, characteristic of Huntington's chorea.

life and far more modest. So you cannot simply equate a gene with a function. Even here, with a single-gene disorder, the gene is not solely responsible for the disease; the environment can intervene and play an important moderating role.

Drugs are another very important area. Drugs work by changing the ability of a chemical transmitter to contact its receptors. So drugs can deplete the release and storage of the transmitter, or they can enhance its release, or they can block the interaction with the receptor, or indeed they can act as an impostor, as with morphine and encephalin. Encephalin is a naturally occurring transmitter for morphine. How does it work? We would say that morphine gives you a dream-like euphoria, but we would not say that euphoria is trapped within a molecule any more than homosexuality is trapped within a gene, or that mini brains are trapped inside a brain region. One of the great challenges in neuroscience is to explain how a molecule can so configure the brain to change our consciousness, and change our state of mind. This is going to be increasingly important in the future when discussions about decriminalising all drugs, including cannabis, are already taking place. It may well be an issue in the clinic too: for example, do we want a whole nation taking Prozac (fluoxetine), even though there is a strong clinical argument for it?

Technology in the future

Finally, we might not like our world and opt instead for the increasingly sophisticated virtual reality and cyber world option. Some teenagers already reject the outside world and opt instead to live in a cyber world. As the software becomes ever more sophisticated, it will become ever more tempting to live there. Will we therefore have not genetic clones but silicon clones, standardised clones? Because if you are exposed to the same software, the same virtual reality all the time, is there not a danger of having standardised human beings? Alternatively, chips could be implanted into the brain, although we will not need implants to learn French because all information will be accessible ubiquitously through invisible computers. I will be able to ask my watch when the Battle of Hastings took place and it will tell me. This, of course, has very interesting moral implications for education and what actually we are learning – that is, differentiating information from knowledge and wisdom.

More immediately for neuroscience, it does suggest that we are close to developing an interface that we have never dreamed of between silicon and carbon systems. One idea that is already working, and a very exciting but sinister one, is the implanting of electrodes which contain nerve growth factor in the brain of a quadriplegic person so that neurones are attracted into interfacing directly with the electrode. This would mean that a paraplegic person could move a cursor on a computer screen by willing it. Just as there are chips in the brain, so the brain or brain cells might soon be found on chips. The scientist Peter Fromhertz has grown on circuit boards neurones which happily talk to the electronic components as easily as they do to each other. We do need a moral perspective which will in the future enable us to evaluate this technology and harness it for good, rather than for evil.

We then have nano science where little 'submarines' will go round your body and report back on your physiological status. This raises interesting moral issues about third-party access to everything to do with your brain and body, including brain scanning. As this area develops, people are for the first time going to have a window on your brain and mind at work, more so than in the past. It is already starting to encroach on our privacy, for good or ill. And if we take nano science to its extreme, we will have unprecedented control over matter by manipulation of atoms. This again raises very serious issues for what Freeman Dyson[4] calls 'neuro technology', which is the manipulation of the state of your mind and the state of your brain processes in a way that we had never envisaged. By the middle of the 21st century, if not before, machines will have surpassed the power of the human brain. That does not mean to say they will be conscious, but in terms of processing power these machines will be infinitely superior to the human brain. I think that is a moral issue, among other things. Some, like Marven Minskey, believe that machines will be conscious. I would suggest, in line with Roger Penrose, that synthetic brains will never be conscious because they do not have, amongst other things, intuition or common sense. Humans work non-algorithmically but machines do not; as the Nobel Laureate Niels Bohr said to a student, admonishingly, 'You are not thinking, you are just being logical.' Computers are logical. As we know from our brain work on chemically based events, drugs work on the chemicals and change our emotions and that is impossible in a computer. Similarly, Stuart Sutherland said that he would believe a computer was conscious if it ran off with his wife, and I have not met any philandering computers. Alan Turing[5] suggested that the distinguishing test for consciousness in a computer is whether a human being, given impartial access to a computer or a person, could not tell the difference in respect of the responses to the questions they asked. Humans sometimes have an inner state which does not necessarily interface with the outside world, does not have to have responses, whereas the whole essence of a computer is that it is something which gives you responses. So I think that the issue of conscious computers is not going to give us insight into how the brain generates consciousness or be particularly valuable to us.

To summarise, the issue will be whether we are going to accept more monitoring of our lives at all levels. Genetic screening and engineering will engender a whole raft of issues that have already started to surface, involving cellular events at the level of one cell to another with both prescribed and proscribed drugs; and brain, body scanning and nano science allowing third-party access to information about your future health. Eventually, issues will arise from heuristic computers and interaction with computers embedded in our bodies and our clothes and the outside environment. This will largely call into question the very boundaries of what we call reality and indeed the boundaries of ourselves. For example, if you and I were all accessing the same cyber space, then where would my mind end and yours begin, if they were interfacing in an agile way across that cyber space? Similarly if I walk into a room and the walls change colour because the sensors in my body say I feel a certain way, what colour are they really? This new technology could have advantages and disadvantages. It could lead to more manipulation and passivity, which we are seeing, or lead to more insight into the human condition, which will throw light on why people take drugs, why people are depressed, why people are schizophrenic and, most important of all, how the brain generates consciousness. It will be our responsibility to influence the way in which future technologies are used, and I think this is really a truly moral issue, and one that should not be left to neuroscience.

References

1. Alkire MT, Pomfrett CJD, Haier RJ et al. Functional brain imaging during anesthesia in humans: effects of halothane on global and regional cerebral glucose metabolism. *Anesthiology* 1999;90/3:701–9.
2. Pascual-Leone A, Nguyer D, Cohen LG et al. Modulation of muscle responses evoked by transcranial magnetic stimulation during the acquisition of new fine motors skills. *J Neurophysiol* 1995;74/3:1037–45.
3. Van Dellen A, Blakemore C, Deacon R et al. Delaying the onset of Huntington's in mice. *Nature* 2000;404:721–2.
4. Dyson F. *Imagined worlds*. London: Harvard University Press, 1997.
5. See for example www.turing.org.uk

2

The neurology of consciousness – and conscience

ADAM ZEMAN

Professor Zeman outlines our knowledge of the biological basis of wakefulness and awareness. He proposes three links between consciousness and morality: moral responsibility is dependent on being conscious; conscience involves self awareness; and consciousness is a major source of moral value. The neurological exploration of consciousness is closely linked to moral concerns.

This chapter aims to sketch what has been learnt over the past century about the neurological basis of consciousness and, at the close, will ask whether the science of consciousness has any relevance to the science of morality – and, in particular, to conscience.

Consciousness as wakefulness

The word 'consciousness' is used in several senses: one of these, especially relevant in medicine, is the sense of wakefulness. While awake, in the conscious state, we are generally conscious of events impinging on us or passing through our minds. This distinction, between our *states* of consciousness, such as wakefulness and sleep, and the *contents* of our consciousness is biologically significant: I will focus first on wakefulness.

The electrical correlates of states of consciousness

The story of the scientific understanding of our states of consciousness has two main strands: the first concerns the electrical correlates of sleep and waking in the brain; the second concerns the structures within the brain which regulate conscious states. This chapter begins by looking at electricity of the brain.

Scientists have known for many centuries that electricity plays some part in the workings of the nervous system. Towards the end of the 19th century it became possible to record the electrical activity evoked by sensory stimuli in the brains of animals.[1] Thus Richard Caton, a professor of physiology in Liverpool working in the 1870s, was able to detect the electrical responses to visual stimuli in the exposed brains of rabbits. He was intrigued, along with his physiologist colleagues elsewhere in Europe, by faint spontaneous currents present in the brains of their experimental animals, but these tiny signals were almost too small to be detected.

It was not until 1929 that spontaneous activity of this kind was first reported in recordings from the human brain – the year in which Hans Berger wrote 'I ... believe that I have discovered the electroencephalogram of man'.[2] Berger was a rather secretive psychiatrist who worked in Jena in Germany at the start of the 20th century, conducting his clinical work by day and his research by night, when the risk of interference from other electrical equipment was minimised. He also kept the hair of his son, Klaus, as short as possible so that he could record from his scalp with a minimum of electrical resistance! His work was inspired by a fascination with the relationship between mind and body, and his main contribution was to show that it is possible to record electrical potentials which reflect our states of consciousness from the scalp.

In Berger's second paper he drew a distinction, which has survived to the present day, between two electrical rhythms which both occur in the waking state (see Figure 2.1). Alpha rhythm, a regular, moderately high-amplitude oscillation at 8–13 cycles/second, can be recorded from the back of the head in a

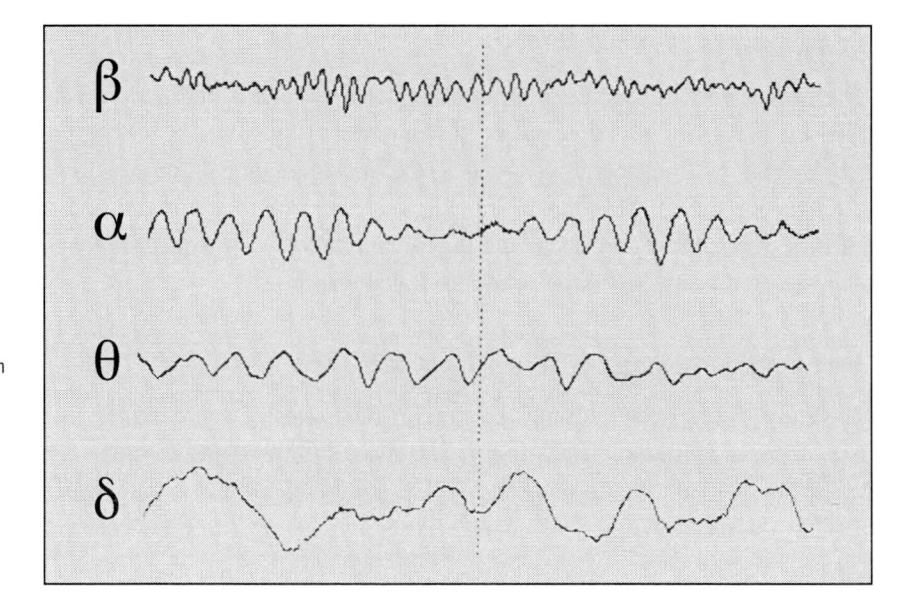

Figure 2.1 The rhythms of the electroencephalogram (EEG). The figure shows two second samples of EEGs from four different patients, exemplifying beta (>14 Hz), alpha (8–13 Hz), theta (4–7 Hz) and delta (<4 Hz) rhythms. Reproduced with permission from Oxford University Press.[3]

relaxed but wakeful subject with the eyes closed; it is the electrical signature of quiet contemplation in the absence of focused mental activity. As soon as the subject engages in a psychologically demanding task, such as mental arithmetic, alpha rhythm is replaced by more rapid (> 13 cycles/second), lower-amplitude activity. Berger christened this 'beta rhythm'. Berger and his followers also found that as wakefulness gives way to sleep, the rapid rhythms of the waking electroencephalogram (EEG) are gradually replaced by slower, higher-amplitude activity – theta (4–8 cycles/second) and delta (< 4 cycles/second).

Today these discoveries are common knowledge and may not seem terribly exciting, but they are of fundamental importance. They demonstrate that states of consciousness have an electrical signature and point to the existence of processes which synchronise activity across the brain. The observation that the diverse regions of the cerebral cortex can work together to generate coherent electrical rhythms, and that these correspond to conscious states, suggests that understanding the integration of brain function is one of the keys to understanding consciousness.

The first clinical application of the EEG was in the diagnosis of epilepsy. Figure 2.2 shows a typical but striking example from a child who is initially awake and aware with correspondingly rapid, low-amplitude EEG rhythms (Figure 2.2). These are abruptly replaced by a burst of the repetitive

Figure 2.2 The electroencephalogram (EEG) during a brief absence seizure.

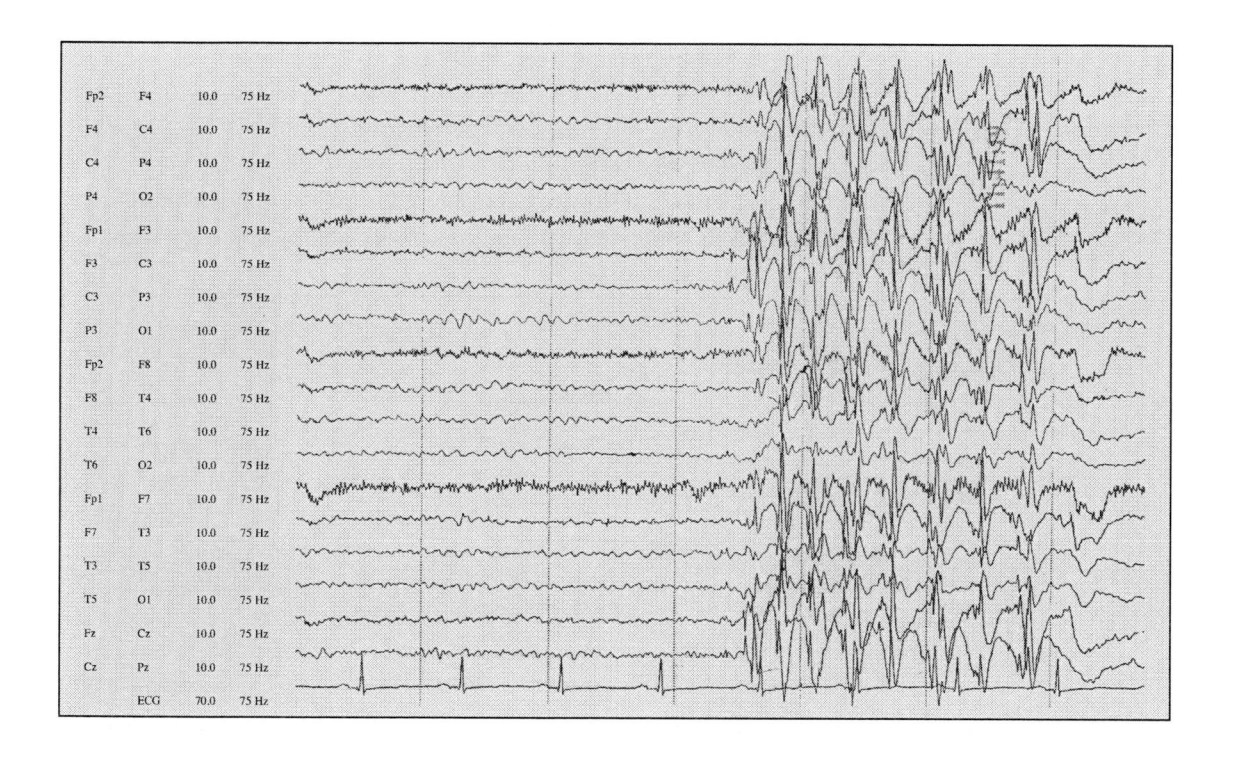

3/second spike and wave activity which characterises absence seizures in children. During an episode of this kind, normal awareness is suspended.

A second application is to the study of sleep and the diagnosis of sleep disorders. In the 1950s the techniques of EEG were applied to sleep with remarkable results. Work from Kleitman's laboratory in Chicago revealed that sleep is not an amorphous process, but rather has a highly organised electrical structure (see Figure 2.3).[4] As we drop off to sleep, the rapid, low-amplitude rhythms of wakefulness are replaced by progressively slower, higher-amplitude activity, until we enter deep slow-wave sleep in which the EEG is dominated by delta waves. Sleeping subjects woken in this stage of sleep will have relatively little mental activity to report (although some mental activity does occur in slow wave sleep).

But, as Kleitman and his colleagues discovered, this phase of deep sleep is transient: after half an hour or so sleepers climb back up the ladder of sleep stages, eventually entering what they called a paradoxical state – paradoxical because, although sleepers are now difficult to wake with profoundly relaxed muscles, their eyes dart here and there beneath their lids and their EEG suggests wakefulness. Woken at this stage of sleep a sleeper is very likely to report a vivid dream. This phase of dreaming sleep is most commonly known as REM – rapid eye movement sleep.

The first period of REM sleep is brief. But over the course of a night the cycle of descent into slow-wave sleep and reascent into REM will repeat itself three or four times. In the first part of the night the periods of slow-wave sleep are prolonged and intense; towards morning REM sleep comes to the fore. This meshes well with ordinary experience: woken at 2 am by a call from the hospital, I find it a considerable effort to haul myself out of deep sleep into wakefulness, and my mind is usually blank; woken by a child at

Figure 2.3 The architecture of sleep: see text for explanation. Reproduced with permission from McGraw-Hill.[5]

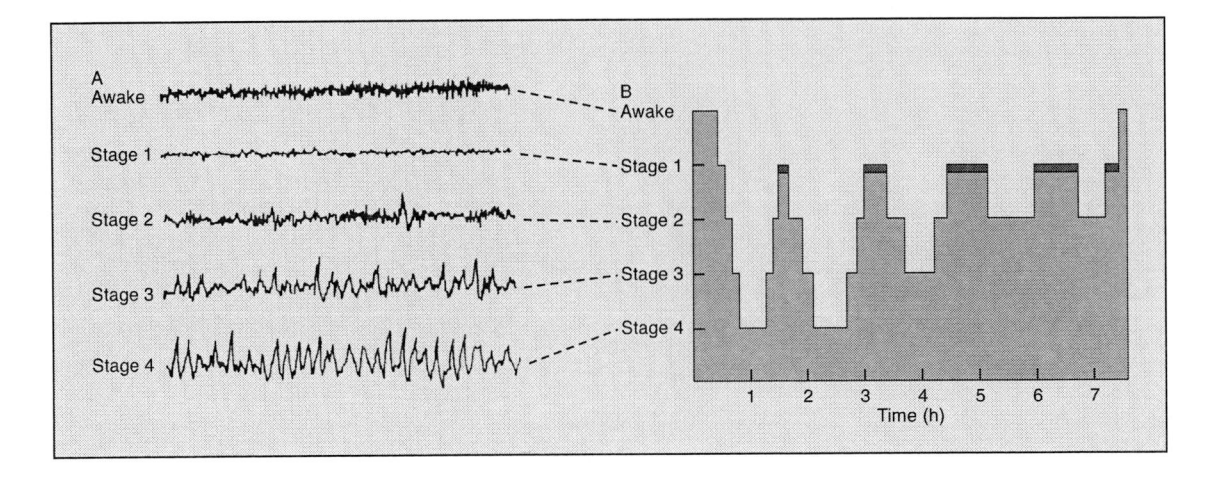

8 am if I have slept in, I am much more likely to find a dream at the forefront of consciousness.

One further, quite recent 'electrical' discovery deserves a mention before we pass on to the brain structures which regulate sleep and wakefulness. Using a modern descendant of the EEG, magnetoencephalography (MEG), Rodolfo Llinas has shown that extremely rapid activity, at around 40 cycles/second, synchronised across the brain, is a feature of wakefulness and REM sleep but not of slow-wave sleep.[6] The idea that so-called gamma rhythms, at 25–100 cycles/second, may play an important part in conscious processes is attracting interest across a broad front at present, from the study of vision to the science of anaesthesia.

The structure of the brain

The second strand in the story of the neurology of conscious states concerns the discovery of the structures which regulate sleep and wakefulness. Towards the end of the First World War a Viennese neurologist and psychiatrist, Constantin Von Economo, was caring for patients with a mysterious encephalitis which has since all but disappeared.[7] Its early features often pointed to a disorder of arousal: some patients presented in a hyperalert, overexcited, hypomanic state; others were in a lethargic, withdrawn, drowsy state. Von Economo recognised that the pathology of the disorder might have something to teach us about the regulation of arousal by the brain, and, as it turned out, he had plenty of opportunity to study its pathology.

Von Economo correlated the clinical features of the illness with findings at post-mortem that suggested that structures in the midbrain, at the top of the brain stem, play a crucial role in maintaining arousal: when these were damaged, patients became drowsy or comatose. He suspected that areas in the hypothalamus played a comparable role in sleep: damage here caused hypomania and insomnia.

At the time of the outbreak of encephalitis lethargica, Frederick Bremner was a young doctor working at the Salpêtrière Hospital in Paris. He became a physiologist, however, and almost 20 years later put Von Economo's ideas to the test in a series of experiments on cats. He showed that transection of a cat's nervous system at the junction between the spinal cord and the brain stem has no impact on the cat's state of arousal or its sleep–wake cycle; by contrast, transection through the upper midbrain causes coma.[8] Like Von Economo, Bremer concluded that structures in the upper midbrain are required to maintain arousal. He suspected that these structures were sensory pathways and that his animals' states of stupor were due to a loss of ascending sensory signals.

Bremer's pupil Giuseppe Moruzzi, working with Horace Magoun, confirmed Bremer's observations but showed that his explanation was wrong.[9] Moruzzi and Magoun identified a region at the core of the upper brain stem which received input from all the senses but whose role was not to transmit sensory signals: instead, this region, which came to be known as the 'reticular activating system', communicates widely with the cerebral hemispheres, directly and via the thalamus, thereby maintaining arousal and regulating our states of consciousness.

Work over the past 30 years has shown that this system is not a single monolithic unit but rather contains a number of chemically distinct subsystems that play different roles in the regulation of conscious states. The chemicals concerned are the neurotransmitters acetylcholine, dopamine, noradrenaline and serotonin, the targets of many of the drugs which act on the brain to modify mood and arousal, among other psychological processes (see Figure 2.4).[10]

❋ ❋ ❋

These two strands of research have explored the electrical correlates of conscious states and the structures in the brain stem and thalamus which control them. This work has helped to define three fundamental states of consciousness in health – wakefulness, and slow-wave and REM sleep – and a number of states of pathologically impaired consciousness.[3,11] The latter will be explored in more detail below: their ethical implications make them relevant to the subject of this book.

Patients in *coma* are neither awake nor asleep. They are in a state of pathological unconsciousness, in which awareness is abolished and the eyes are closed. Levels of cerebral metabolism are markedly, but variably, reduced. The state is almost always transient – in other words the patient is en route to recovery, death or the vegetative state.

The *vegetative state* is a condition of wakefulness without awareness, one of the outcomes of coma, in which cycles of sleep and waking occur in the absence of any good evidence of a functioning mind.[12] The eerie dissociation between apparent wakefulness and awareness can occur because the structures regulating the sleep–wake cycle in the brain stem, as already discussed, are distinct from those in the hemispheres mediating the contents of awareness, as discussed below. Cerebral metabolism is very markedly depressed in the vegetative state, down to levels seen under deep general anaesthesia.[13] Recovery from the vegetative state can occur. It is deemed persistent if it lasts for more than one month and permanent if the prospects of recovery are judged to be negligible.

The *minimally responsive state* is often seen in the course of recovery from the vegetative state and is sometimes its final outcome. This is a state of low

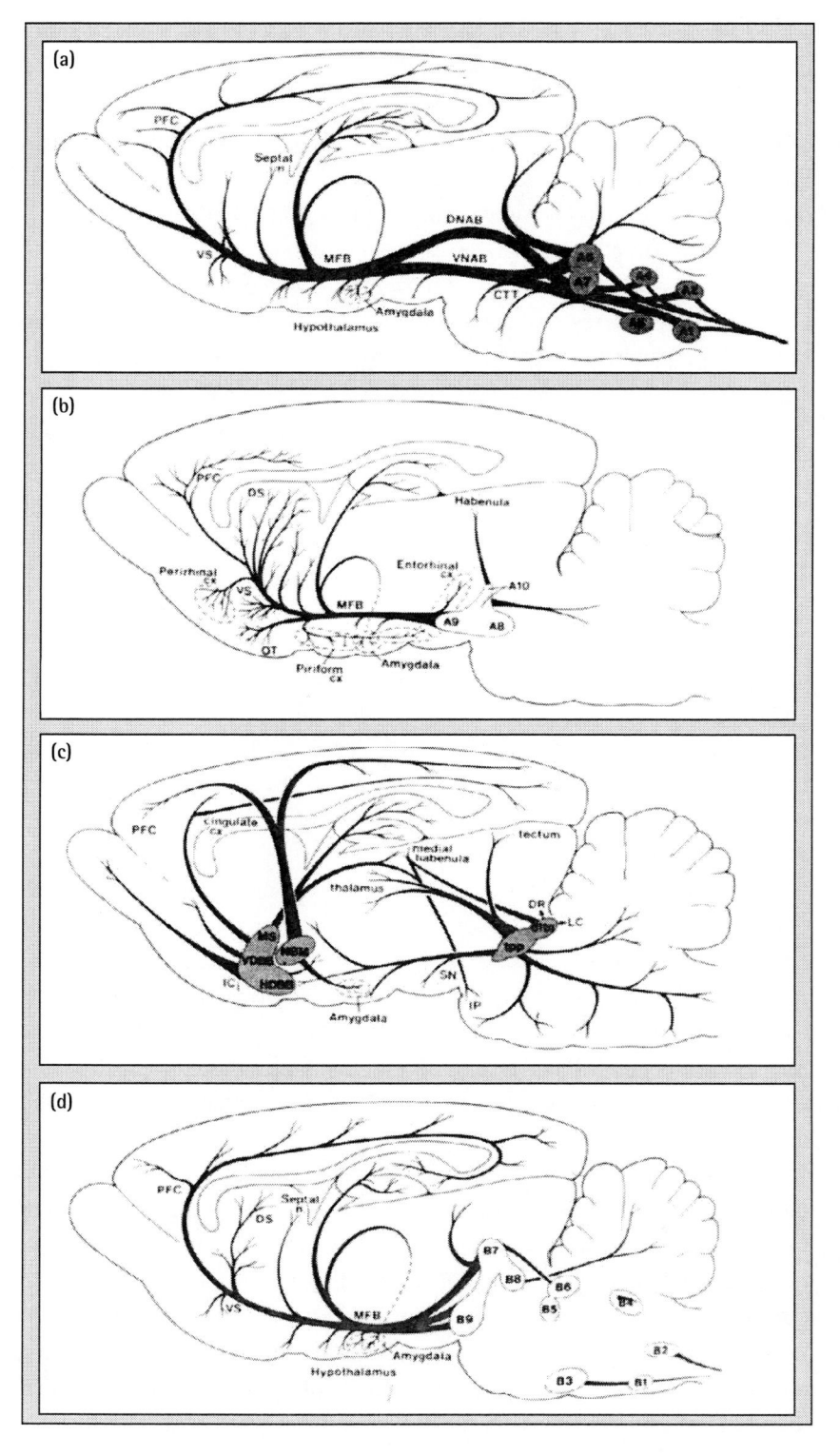

2 The neurology of consciousness – and conscience

Figure 2.4 The chemistry of consciousness: this figure shows chemically distinct components of the ascending activating system, transmitting noradrenaline (a), dopamine (b), acetylcholine (c) and serotonin (d). CTT = central tegmental tract; dltn = dorsolateral tegmental nucleus; DNAB = dorsal noradrenergic ascending bundle; DR = dorsal raphe; DS = dorsal striatum; HDBB = horizontal limb nucleus of the diagonal band of Broca; Icj = island of Calleja; IP = interpeduncular nucleus; LC = locus ceruleus; MFB = medial forebrain bundle; MS = medium septum; NBM = nucleus basalis magnocellularis (Meynert in primates); OT = olfactory tubercle; PFC = prefrontal cortex; SN = substantia nigra; tpp = tegmental pedunculopontine nucleus; VDBB = vertical limb nucleus of the diagonal bond of Broca; VNAB = ventral noradrenergic ascending bundle; VS = ventral striatum. Reproduced with permission from MIT Press.[10]

awareness associated with consistent or inconsistent evidence of awareness in the presence of major impairment of cognitive function.

In the *locked-in* state, patients are both awake and aware but unable to communicate normally because of damage to the brain stem interrupting the bulk of descending signals to the head and the limbs. Vertical eye and eyelid movements are preserved and provide a limited channel of communication.

Brain death is defined in the UK as death of the brain stem, abolishing the sleep–wake cycle and awareness, with irrevocable loss of spontaneous ventilation and all the reflexes mediated by the brain stem. It is usually followed by cardiac death within hours or days. It has great ethical and legal significance, as organ recovery is permitted once this diagnosis has been established, assuming the patient or his or her relative has given the appropriate consent.

This chapter has concentrated so far on the neurology of our states of consciousness. We will now turn to the contents of consciousness and consider what can be learnt from science about awareness.

Consciousness as awareness

Imagine the world as it would appear to someone who had suddenly lost the ability to see in colour – all the familiar refreshing hues of the visual world reduced to shades of grey. This bizarre phenomenon occasionally occurs, usually on just one or other side of visual space, after a stroke. The disorder is called central achromatopsia (central because the problem lies in the brain and not in the eye).

One of Oliver Sacks' wonderful accounts of neurological disorders, the case of the colour-blind painter, outdoes the standard descriptions of central colour blindness.[14] The subject of this essay is an American painter, Jonathan I, who specialised in beautiful, highly coloured abstract canvases. He abruptly lost colour vision right across his field of vision following a head injury, and he became very depressed as a result, but eventually he adjusted to his loss and tried to convey his changed experience of the world through a series of severe monochromatic paintings.

What is the explanation for this extraordinary disorder? How is it possible for damage to the brain to abolish the consciousness of only a single aspect of the visual world? An answer to this question has emerged from work on vision in animals and humans over the past century. It has been known since Richard Caton's time that visual stimuli excite electrical activity in the brain. In humans the strongest responses are in the occipital lobe at the back of the brain. During the First World War the British neurologist Gordon Holmes established that the occipital cortex mapped the field of vision – ie there was a predictable relationship between the region of damage to the occipital lobe in injured soldiers and the region of the visual field in which they had lost vision.[15]

More recent work has shown that the map Gordon Holmes plotted in the primary visual cortex of the occipital lobe, labelled V1, is only one of around 30 maps of the visual world in areas which surround V1, extending into the parietal and temporal lobes.[16] Why should there be so many and what do they do? At least part of the answer is that these areas share out the computational work of vision, specialising in different aspects of the neural processing that enables us to see.

Semir Zeki, a physiologist at University College, London, has highlighted a particularly clear-cut contrast between two of these visual areas, V4 and V5.[17] Using the techniques of functional imaging, which make it possible to demonstrate the activity in brain regions associated with particular psychological tasks, Zeki has shown that area V4 plays a key role in the perception of colour, while V5 plays a comparable role in the perception of visual motion. This is not to imply that activity in V4 is sufficient for the perception of colour – Susan Greenfield's chapter gives some good reasons to question claims of this kind – but it does at least appear to be necessary for the experience. And, of course, the close correlation between activity in a particular region of the brain and a certain kind of visual experience helps to explain what was puzzling initially: how the kind of local brain damage caused by a stroke can abolish something as specific as the conscious perception of colour.

Figure 2.5 is a sketch of the visual areas and some of their interconnections and indicates the complexity of the visual brain. But it also helps to draw attention to a helpful simplification, for it seems that there are two major streams of visual information flowing through these areas. One, the ventral stream (or 'what' pathway), flows from area V1 into regions in the temporal lobe which are particularly concerned with colour perception and object identification (area V4 belongs to this stream); the other dorsal (or 'where' pathway) flows up into the parietal lobe and is particularly involved in the visual guidance of action.

Work along these lines demonstrates close – sometimes exquisitely close – correlations between aspects of visual experience and aspects of the activity of the visual brain. But there is a problem lurking here. When you show a visual stimulus to a subject all kinds of things happen in the brain as a result. Some of these are likely to be related to the resulting visual experience, but others are not. For example, if a light shines in your eye, your pupil will rapidly contract as a result of the functioning of a reflex arc which passes through the brain but which has very little to do with consciousness: the pupil will constrict quite normally in someone rendered completely blind by

Figure 2.5 The cortical visual areas: the sketch shows some of the 30 cortical visual areas. Areas in the temporal 'what' pathway are shaded. TF lies in the parahippocampal gyrus. Area PG includes area 7a, a subdivision of Brodmann's area 7. PO = parieto-occipital; VIP = ventral intraparietal; MT = middle temporal; MST = medial superior temporal; FST = floor of superior temporal; VOT = ventral occipito-temporal; PIT = posterior inferotemporal; CIT = central inferotemporal; AIT = anterior inferotemporal; FEF = frontal eye fields. Reproduced with permission from Elsevier.[16]

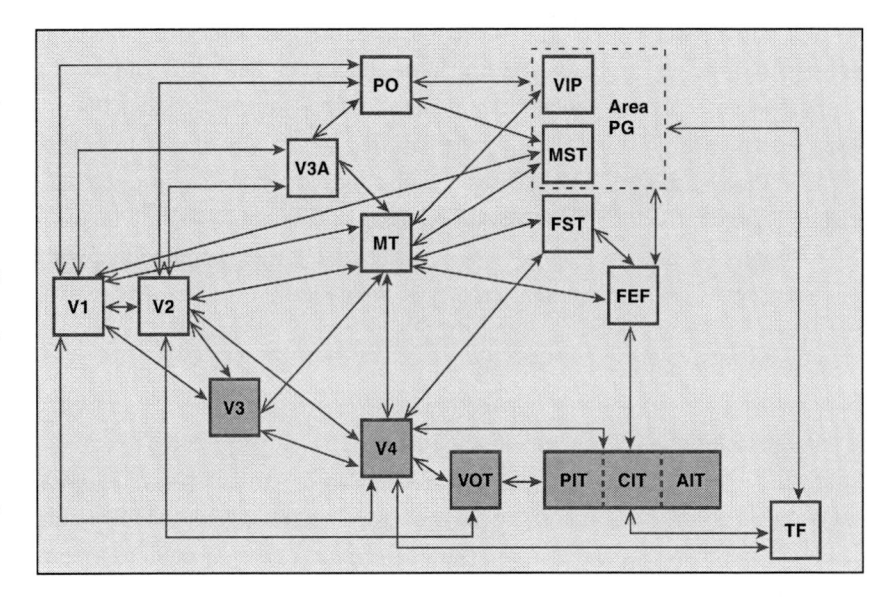

damage to V1. If we are keen to home in on the varieties of neural activity which give rise to consciousness, we need to find some way of distinguishing these from the unconscious processes which normally accompany them. There are at least two ways of doing this.

The first is to try to identify the activity in the brain which correlates with changes of experience occurring in the absence of changes in the world outside us. This sounds odd in the abstract but the idea is simple: for example, if you close your eyes, imagine your kitchen and ask yourself where you would look for the jam, something is happening in your mind's eye, and presumably in your brain, without any change in the world outside your head. If one could pick out the activity which correlates with this act of visual imagination, it should be tied fairly closely to the neurology of conscious experience. A number of scientists have followed up this line of thought.

The beautiful work of an American neuroscientist, Nancy Kanwisher, provides a good example.[19] She has worked with two areas well down the stream of visual processing in the brain. One, the fusiform face area (FFA), is strongly activated by images of faces; the other, the parahippocampal place area (PPA), is strongly activated by images of physical locations, such as houses. The existence of these areas is interesting in itself, but of course activity in them might not be directly tied to the conscious perception of places and faces. Kanwisher has specifically investigated this question.

Using functional imaging, she has shown, first, that the areas are activated by *imagining* a face or place: the fact that an external stimulus is not required supports the idea that the areas may be involved in conscious perception.

Second, she has shown that when the eye is simultaneously presented with images of faces and places above and to the side of the point of fixation, activity increases in the FFA when attention is directed to the faces and in the PPA when attention is directed to the places. Third, she has shown that activity in the PPA and FFA keeps time with the alternation of visual percept under conditions of binocular rivalry: this requires a little explanation!

If the two eyes are presented with different images, one with the image of a face, the other with the image of a house, what we see tends to be not a fusion of the two but an alternation between them: for a second or two the face, then the house, then back to the face, and so on. Work in animals has shown that this alternation is not apparent early in the visual pathway: thus in area V1, cells will be responding to both stimuli all the time. However, if subjects are asked to indicate the moment at which the images alternate, and activity in the PPA and FFA is tracked using functional imaging, the neural activity turns out to mirror the alternation in the perceived image.

Kanwisher's work suggests that the FFA and PPA are parts, at least, of the neural system which underlies the conscious perception of faces and places. Another nice example of the work pursuing this line of attack comes from Dominic Ffytche at the Institute of Psychiatry in London,[20] who has studied patients with the Charles Bonnet syndrome. This term describes the occurrence of visual hallucinations in patients who lose sight late in life, usually from pathology affecting the eyes rather than the brain. Hallucinations are common among this group, although sufferers are often wary about discussing them for fear that others will think they have gone mad. They have not gone mad: they are perfectly well aware that the hallucinations, which come and go, are unreal, and they are not, as a rule, too concerned about them. Because the hallucinations are intermittent, Ffytche was able to ask his subjects to indicate their onset and offset and to compare brain activity while they were present and when they were absent. This project revealed activity in higher visual areas in the region of the FFA and PPA. It was possible, to some extent, to correlate the content of the hallucinations with the location of the active visual areas. Ffytche also showed that activity in these regions built up over a second or so before his subjects reported the onset of a hallucination.

The approach taken in the work of Nancy Kanwisher and Dominic Ffytche helps to consolidate the link between certain kinds of brain activity and certain types of conscious experience.

It was mentioned earlier that there is a second approach to the challenge of distinguishing conscious from unconscious processes in the brain. It takes a quite different tack. The guiding idea is that if much of the activity occurring in the brain occurs without exciting consciousness, clarifying the location and nature of these unconscious processes should help to highlight the distinctive qualities of conscious processing.

The most intensively studied unconscious process of this kind is probably blindsight. This is a fascinating story which has recently been brought full circle. It was known in the 1970s that after damage to the primary visual cortex in monkeys, visual function eventually recovers rather well: in other words, the monkeys become able to guide their movements using vision despite damage to the corresponding region of V1. This was a puzzle, because damage to V1 in humans appeared to cause complete and permanent blindness. The Oxford psychologist Larry Weiskrantz wanted to reconcile these two apparently conflicting observations. He had the extremely simple but inspired idea of inviting a patient with selective damage to V1 to guess what might be happening in his blind field. To Weiskrantz's delight and the patient's amazement these guesses were usually correct: this patient, and a number of others tested since, were able to indicate the presence or absence, the direction of movement, the orientation and the shape of visual stimuli which he was quite unable to see.[21] This suggests that quite sophisticated visual processes can occur in the absence of awareness.

Weiskrantz's colleague in Oxford, Alan Cowey, and the German neuropsychologist Petra Stoerig have posed the question: could monkeys who recover visual function after damage to V1 also be using blindsight? They devised an ingenious experiment.[22] After establishing that the monkeys could detect stimuli in their blind fields very reliably, Cowey and Stoerig trained the monkeys to respond to stimuli in their *sighted* fields in an experiment in which some trials contained *no* stimulus. In these cases the monkeys had been trained to press a special panel which meant that no light had been seen. While the monkeys were performing this task, Cowey and Stoerig slipped in an occasional flash in the monkey's blind field. On these trials the monkeys continued to press the no-stimulus panel, although the previous experiments had shown that they were perfectly well able to detect a stimulus of this intensity in the blind field. In contrast, a control subject, Rosie, with normal vision, immediately pressed the panel which had flashed, to whichever side it occurred. Cowey and Stoerig concluded that after damage to V1, monkeys can detect stimuli in their blind fields, but these stimuli no longer have visual qualities – just as in humans.

On first encounter blindsight can seem totally mysterious. It is not: the royal road from the retina to the thalamus and on to the primary visual cortex is only one of many routes taken by visual information from eye to brain: there are numerous subcortical and a few other cortical destinations (Figure 2.6).[23] Which of the alternative routes subserves blindsight is uncertain, but there are plenty of potential candidates.

2 The neurology of consciousness – and conscience

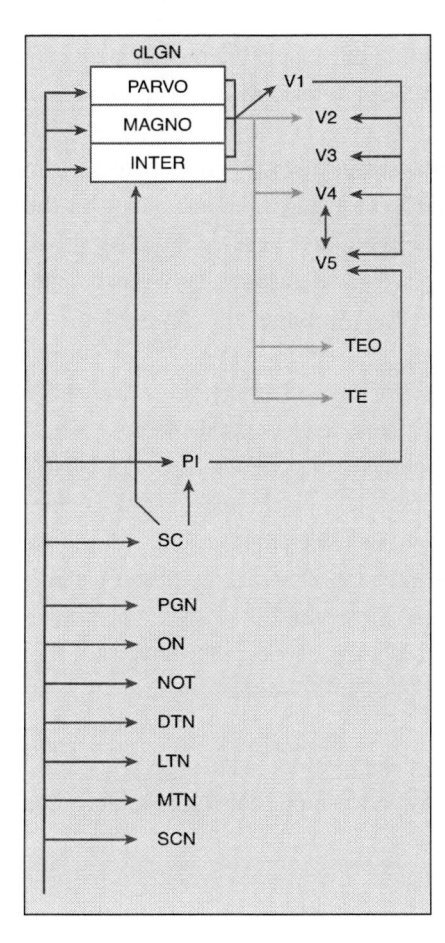

Figure 2.6 Projections from the retina: dLGN is the lateral geniculate nucleus, main recipient of visual signals from the retina. But signals also travel to the superior colliculus (SC), the pregeniculate nucleus (PGN), the olivary nucleus (ON), the nucleus of the optic tract (NOT), the medial, lateral and dorsal terminal accessory nuclei (MTN, LTN, DTN) and the suprachiasmatic nucleus (SCN). Pl is the pulvinar nucleus. There are relatively sparse direct projections from dLGN to visual cortical areas other than V1 (such as V2, V4, FEO, TE). PARVO = parvicellular; MAGNO = magnocellular; INTER = interlaminar. Reproduced with permission from Oxford University Press.[23]

I have shown that there are close correlations between activity in the brain and the characteristics of visual experience, and that there are at least two approaches which enable one to focus down on the neurology of awareness: by picking out activity in the brain which changes when experience changes in the absence of stimulus change (which are likely to be particularly central to awareness), and by identifying the activity which subserves unconscious processes (which are, therefore, shown to be insufficient for awareness).

Before examining the links between consciousness and conscience, I will explain a little about the kinds of overarching theories of consciousness that have been proposed on the basis of this evidence.

Theories of consciousness

A number of theories of consciousness are on offer in neuroscience at present, but there are several broad points of consensus. It is widely agreed that consciousness arises from the brain but that only some brain events are conscious; that our states of consciousness are determined mainly by processes in the brain stem and the thalamus, while the contents of consciousness depend upon activity in the cerebral hemispheres, especially the cerebral cortex; and that the kinds of neural activity on which consciousness depends are spread around the brain, in shifting coalitions between nerve cells, sometimes called cell assemblies. Most theories also agree that interactions between psychological processes, such as memory, action planning and perception, are required for consciousness.

Francis Crick and Gerald Edelman, two Nobel Prize winners who have turned to consciousness in their later careers, have devised theories of this general type.[24,25] (See also Susan Greenfield, this volume, pp1-11.) Larry Weiskrantz, the psychologist who put blindsight on the map, paints a comparable picture with a psychological emphasis.[26] He proposes that direct links between visual and motor areas of the brain subserve the abilities of patients with blindsight: *conscious* vision requires the brain to be capable of providing a further commentary upon these processes. Weiskrantz envisages that this emerges from areas of the brain in the limbic system and frontal lobes linked to memory and action.

Thus the view of consciousness emerging from contemporary neuroscience is that it results from interactions between multiple brain regions and psychological processes which individually are unconscious.

All these theories are open to the question of why *any* interactions of these kinds should give rise to *consciousness* (as opposed to complex behaviour). This is sometimes called the hard problem of consciousness,[27] and it will not be explored further here. The following section examines some possible links between consciousness, conscience and morality.

Morality, consciousness and conscience

Consciousness as a prerequisite for responsibility

The first link is fairly straightforward. Someone would not normally be regarded as responsible for an action unless they were conscious at the time that they performed it. Here is an example:

> *I was on a motorcycle going down the highway when another motorcyclist comes up alongside me and tries to ram me with his motorcycle. Well, I decided I'm going to kick his motorcycle away and at that point my wife woke up and said "What in heavens are you doing to me?" because I was kicking the hell out of her. In the dream he saw clearly, heard nothing and felt fear of being rammed.*

This account was given by a 'dapper, pleasant, well-adjusted man aged 67 enjoying retirement'. He had always been a rather restless sleeper and began to act out his dreams at the age of 63, at considerable risk to his own and his wife's health.[28] His condition, first recognised about 20 years ago, is REM sleep behaviour disorder: patients with this disorder fail to develop the atonia of REM sleep which normally prevents us from acting out our dreams – with potentially serious results for sleeping partners. One would probably not want to hold this patient responsible for an act committed in his sleep.

Here is another example:

A man was aroused early at night by his wife, who was shouting a false alarm of 'There are robbers, burglars in the backyard'. In a confused state the husband impulsively grabbed a gun from his night table, ran to the front window and killed a night watchman in the street.[29]

This act of homicide resulted from 'sleep drunkenness', a form of confused arousal from deep sleep.

Responsibility for a crime requires both the commission of a culpable act, *actus reus*, and the existence of a culpable intention, *mens rea*. Framing a culpable intention requires that you are conscious. An act performed while unconscious is technically described as an automatism. If the cause of the automatism is an internal state which is likely to recur, the automatism is regarded as insane. The insanity defence is governed by the McNaughton Rule: for the defence to be successful it must be shown that at the time of committing the act in question, the accused was 'labouring from such a defect of reason from disease of the mind as not to know the nature or quality of the act he was doing, or, if he did know it, as not to know that what he was doing was wrong'.

In the two examples above the actors' states of altered awareness arguably prevented them from knowing the 'nature or quality' of their acts. The second element of the McNaughton Rule indicates that you might also be exonerated from responsibility if you did not know that what you were doing was wrong. This points us towards conscience.

The place of conscience in human consciousness

Think about the last time you did something wrong. Our awareness of wrongdoing tends to dominate consciousness – for a little while, at least. This awareness has two major elements. First, the ability to acknowledge that you have voluntarily performed some act, and, second, the ability to judge that it was wrong. The first requires a degree of self-awareness, knowledge of oneself as a free agent. The second requires that one subscribes to a code of conduct.

These two sets of abilities – self-awareness and the possession of a moral code – are major ingredients of human consciousness, but they are not synonymous with it. I think one can imagine people who are conscious but neither able to represent to themselves the fact that they have performed an act nor able to judge that it was wrong: a combination of damage to the limbic system and the frontal lobes, for example, could give rise to such a state of affairs.

It is striking, though, that our culture has long perceived close links between conscience and human consciousness. Here is a passage from the Bible:

So when the woman saw that the tree ... was to be desired to make one wise, she took of its fruit and ate... Then the eyes of both were opened, and they knew that they were naked; and they sewed fig leaves together and made themselves aprons.[30]

The Book of Genesis draws a clear link between the getting of full human consciousness, including the knowledge of good and evil, and the arrival of conscience, guilt and shame.

Charles Darwin makes a similar connection, in a more generous spirit:

any animal whatever, endowed with well-marked social instincts, would inevitably acquire a moral sense or conscience, as soon as its intellectual powers had become as well developed, or anything like as well developed, as in man.[31]

In Darwin's view, evolving human intelligence, in combination with human sociability, were bound to lead 'to the golden rule, "As ye would that men should do unto you, do ye to them likewise", and this lies at the foundation of morality'.

Consciousness as the source of moral value

The final link between consciousness and morality is the most fundamental, and perhaps also rather obvious. Most moral theories, though not all, regard conscious beings as the main focus of moral concern. If this is accepted, the science of consciousness becomes quite important for morality as it will play a part in helping us to decide who or what is or is not conscious: This dog? That spider? The robot that will be at your service in 25 years? The alien intelligence we may soon be encountering?

What we already know for sure is that recognising consciousness in others can be extremely difficult. Take this example:

I did not know who or where I was, or what on earth was happening... This relatively happy state was interrupted by a voice in the space above me (some remark about my bladder) and I instantly understood my predicament: that I was lying ... covered in green towels, my abdomen split open... l remained in this state of mind ... continuously filled with fear, listening to every word, every sound in the theatre, quite compos mentis and fully appreciating my position... The pain ... was bad from the onset and it increased in severity... The nearest comparison would be the pain of a tooth drilled without local anesthetic – when the drill hits a nerve. Multiply this pain ... l can even feel the breath of it now as I am writing all this down.[32]

These moving words are taken from the 'unedited recollections of a medically qualified lady' who had the misfortune to regain awareness during a caesarean. Her description of her experience was published, as a salutary tale, in the *British Journal of Anaesthesia*. It is of course very difficult for anaesthetists to know whether patients are conscious because they often paralyse them to ease the surgeon's task. As a result, anaesthetists are working hard to find reliable indices of awareness using sophisticated versions of the EEG, with some success.

❉   ❉   ❉

Science does then, it seems, have something to say about consciousness – and a little to say about conscience. To make the most of science in the domain of ethics, scientists, clinicians and philosophers will need to join forces over the coming century.

References

1. Brazier M. *A history of the electrical activity of the brain: the first half-century*. London: Pitman, 1961.
2. Gloor P. *Hans Berger on the EEG of man*. New York: Elsevier, 1969.
3. Zeman A. Consciousness. *Brain* 2001;124:1263–89.
4. Dement W, Kleitman N. Cyclic variations in EEG during sleep and their relation to eye movements, body motility and dreaming. *Electroencephalogr Clin Neurophysiol* 1957;9:673–90.
5. Kandel ER *et al*. *Principles of neural science*. East Norwalk, Connecticut: McGraw-Hill, 1991.
6. Llinas R, Ribary U. Coherent 40-Hz oscillation characterises dream state in humans. *Proc Natl Acad Sci USA* 1993;90:2078–81.
7. Von Economo C. *Encephalitis lethargica: its sequelae and treatment*. Oxford: Oxford University Press, 1931.
8. Bremer F. Cerveau isole et physiologie du sommeil. *Comptes Rendus de la Societé de Biologie* 1929;10:1235–41.

9 Moruzzi G, Magoun HW. Brain stem reticular formation and the activation of the EEG. *Electroencephalogr Clin Neurophysiol* 1949;1:455-73.
10 Robbins TW, Everitt BJ. Arosal systems and inattention. In: Gazzaniga MS (ed), *The cognitive neurosciences*. Cambridge, MA: MIT Press, 1995.
11 Zeman A. *Consciousness: a user's guide*. London: Yale University Press, 2002.
12 Jennett B, Plum F. Persistent vegetative state after brain damage. *Lancet* 1972:1: 734-7.
13 Laureys S, Berre J, Goldman S. Cerebral function in coma, vegetative state, minimally conscious state, locked-in syndrome, and brain death. *Yearbook of Intensive Care and Emergency Medicine* 2001:386-96.
14 Sacks O. *The anthropologist on Mars*. London: Picador, 1995.
15 Holmes G, Lister WT. Disturbances of vision from cerebral lesions, with special reference to the cortical representation of the macula. *Brain* 1916;39:34-73.
16 Douglas RJ, Martin KA, Nelson JC. The neurobiology of primate vision. In: Kennard C (ed), *Visual perceptual defects*. London: Baillière Tindall, 1993.
17 Zeki S. *A vision of the brain*. Oxford: Blackwell Scientific, 1993.
18 Kanwisher N. Neural events and perceptual awareness. *Cognition* 2001;79:89-113.
19 Ffytche DH, Howard RJ, Brammer MJ, *et al*. The anatomy of conscious vision: an fMRI study of visual hallucinations. *Nat Neurosci* 1998;42:19-24.
20 Weiskrantz L. *Blindsight: a case study and implications*. Oxford: Clarendon Press, 1998.
21 Cowey A, Stoerig P. Blindsight in monkeys. *Nature* 1995;373:247-9.
22 Stoerig P, Cowey A. Blindsight in man and monkey. *Brain* 1997;120:535-59.
23 Edelman GM. *Bright air, brilliant fire*. London: Penguin Books, 1992.
24 Crick F. *The astonishing hypothesis*. London: Simon and Schuster, 1995.
25 Weiskrantz L. *Consciousness lost and found*. Oxford: Oxford University Press, 1997.
26 Chalmers DJ. *The conscious mind*. Oxford: Oxford University Press, 1996.
27 Schenck CH, Bundlie SR, Ettinger MG *et al*. Chronic behavioural disorders of human REM sleep: a new category of parasomnia. *Sleep* 1986;9:293-308.
28 Bonkalo A. Impulsive acts and confusional states during incomplete arousal from sleep: crimunological and forensic implications. *Psych Quart* 1974;48:400-409.
29 Genesis 3:6-7. Revised Standard Version.
30 Darwin C. *The descent of man, and selection in relation to sex*. London: John Murray, 1871.
31 Anonymous. On being aware. *Br J Anaesth* 1979;51:711-12.

3

The evolutionary genetics of morality

IAN CRAIG AND CAROLINE LOAT

The question of the impact of pre-programmed behaviour is considered in several chapters. Professor Craig and Caroline Loat introduce the evidence for genetic determinants. The rapid progress in psycho-genetics since the definition of the human genome has changed the balance of the nature versus nurture discussion. It was surprising to find not 150,000 genes, as had been predicted, but 30,000, only about twice as many as those of a fruit fly. The sophistication of the human being is not explained by the determinist view of one gene per character - there are too few - or by the reductionist view that understanding the genome will lead to a complete explanation of human variability. Variability lies in the subtlety of control genes and the contribution of many genes of small effect to the overall character. Nerve transmission, for example, involves the genetic control for the production, transport, reception, destruction and inhibition of neurotransmitters such as serotonin. The discovery of large numbers of genetic variants called single nucleotide polymorphisms (SNPs - pronounced snips) also has helped to solve the conundrum of how human sophistication and variability has evolved with so few genes. These findings raise the ethical issue of punishment for antisocial behaviour which may be of genetic aetiology. The known function of genes for neurotransmitter release and detection is contrasted with the genetic determinants for civility, intelligence and morality.

This chapter is about genetics and is an attempt to see where genes and evolution fit into the overall picture of the development of morality. The thesis is simply that morality is an evolutionary mechanism with survival implications. If this is the case, then morality will have genetic as well as environmental inputs. For some people, it may seem that morality should also be subject to additional influences such as spirituality, but for the time being we will concentrate on the interplay between nature and nurture. Research in behavioural genetics concentrates on searching for genes involved in complex conditions, such as conduct disorders and deficits in cognition. Morality is a similar kind of complex trait and as such should be amenable to the same sort of analysis that is used in studies on, say, aggression or intelligence.

We will begin by analysing moral development in a basic way and looking within its aetiology for evidence of a genetic or environmental effect. In particular, it is instructive to look at potential evolutionary explanations for behavioural patterns, which may interact with, or support, what we understand as morality. For this we will need to draw on the extensive work of experimental behaviourists.

The potential explanations for morality can then be examined in terms of 'hard science', and genetic analysis can be applied to try to identify specific genes affecting behaviours, such as aggression and antisocial behaviour, which appear to have moral significance. Contemporary researchers are not doing anything particularly new in looking for genetic influences on behaviour and Figure 3.1 is a useful reminder of this fact. This figure shows an example of an understanding of the potential importance of heredity from the admission sheet to the Bethlem Royal Hospital front-sheet dated 1823; the person in question was admitted as being 'dangerous to others', but it is also noted that he has a brother who is dangerous too. So genetics has for a long time been part of the framework for considering the aetiology of antisocial (ie non-moralistic) behaviour

The development of moral concepts in childhood and their evolutionary significance

In the search for evidence that moral behaviour is an evolutionary mechanism, we need to look at how it develops as children grow up. We can start by examining Piaget's theory of moral development.[1] He described young children as behaving in a totally egocentric fashion. He further considered that as time progressed, they become more sophisticated and aware of other people's perspectives. Piaget was probably over-simplifying the situation by defining it in rather broad terms.

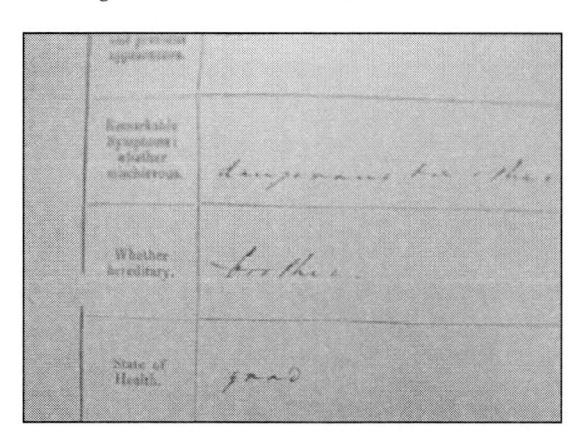

Figure 3.1 Excerpt from a Bethlem Royal Hospital front-sheet dated 1823, courtesy of Professor Peter McGuffin. The entry log records the patient as being 'dangerous to others'. It further demonstrates an understanding of the possible hereditary nature of the behavioural disorder by stating a brother is (presumably) similarly afflicted.

Later, though, Kohlberg took Piaget's framework as a basis for a more detailed breakdown of moral development and the stages involved.[2] First of all he described stages 1 and 2, which make up the so-called 'preconventional level' where children are basically egocentric and unable to consider other people's feelings. They act primarily for the satisfaction of their own desires and avoidance of punishment, but with some orientation towards pragmatic exchange and reciprocity. Then, as children mature, they progress to stages 3 and 4, where the emphasis shifts, and pleasing and gaining approval from others become important. This requires children to have an understanding of theory of mind and the way in which other people interact with them. They begin to define what is right in terms of what is expected by their close family and their identified stereotypes. This then extends from the family to apply more widely in society, and moral judgements are made according to a regard for one's duty and the social order. Individuals may progress further in their moral reasoning, depending upon a variety of factors such as whether they are able to construct a social contract using the principles underlying morality. Finally, at a level achieved by only a minority of people, there is a kind of transcendental use of upper moral structure, which is irrespective of what society believes. This is a moral construct that the individual works out for himself, and may differ from what is dictated as right by societal laws. This overall structure can be refined even further, as has been done by Elliot Turiel for example, who distinguished between morals and conventions.[3] It is interesting to note, however, that most considerations of this sort do not distinguish moral development in boys from that in girls. This seems to be an important area worthy of future thought.

So what evidence is there that there may be a genetic programme working away in the background of moral development? The first aspect relevant to this is in an area that could be referred to as 'soft' science and stems from the concept that there is essentially a similarity of mental organs and their development across all population groups (a concept that has almost universal acceptance). As Steven Pinker said, 'People of the world share an astonishingly detailed universal psychology.'[4]

Hence, there is a concept of morality and brain development that extends across culture and geographic regions. There is a commonality of psychology, which suggests that there is an inbuilt genetic direction operating. Pinker[4] also pointed out that identical twins share much of the fine structure of their personality and intelligence, even when they are brought up apart. Of course, in many ways, twin studies underpin everything we know about behaviour, in that they have allowed the relative contributions of genetics and the environment to be partitioned. Let us, then, examine behaviours that appear to have a familial concentration and see how the study of pedigrees can help in trying to decide how much of a given behavioural trait is due to genetics.

If we take very strongly genetically determined behaviours such as the mental disturbances associated with Huntington's disease, we find of course that there is a very firm familial correlation. We find the same, but to a much lesser extent, with manic depression and schizophrenia, and also intelligence. Simply using the criterion of familial correlation, however, we could also select traits such as religion and allegiance to a football team as being genetically determined. So just because something is familial, it does not automatically mean that it is genetic. In fact, it turns out that attending medical school shows a strong familial inheritance pattern, and if we analyse it in more detail then we find that the risk factor for attending medical school is 80 times greater for your siblings if you attended medical school yourself. It appears to segregate as an autosomal recessive disorder in the population![5]

But, flippancy apart, studies on family relationships are a major approach to gauging the amount of genetic input into behaviours and, in the same way, such approaches may allow one to ask questions about the aetiology of morality. We will consider this aspect in more depth later.

Figure 3.2 provides a good example of family studies providing evidence for genetic influence on a complex disorder (schizophrenia). As the proportion of genes shared between individuals increases, the relative risk of the disorder increases from a very low figure of under 1% for the general population, where individuals taken at random do not share many genes, to a risk as high as 50% in identical twins, where essentially all genetic material is shared between the twins.

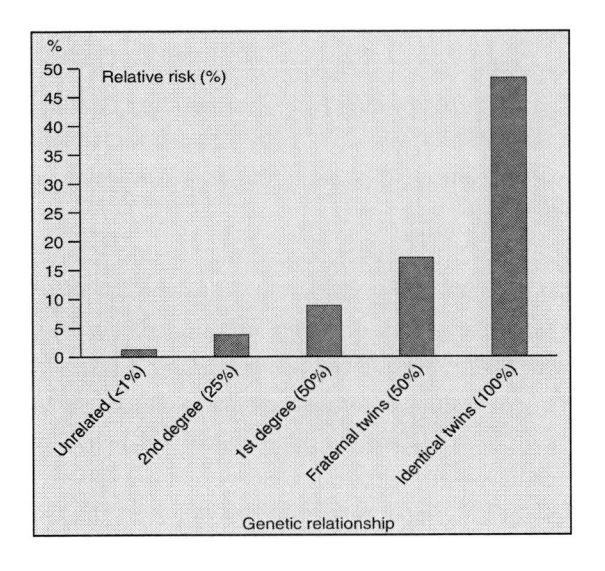

Figure 3.2 Genetic relationships and risk of schizophrenia. The graph illustrates that the relative risk of schizophrenia is about 7% for siblings, close to 20% for fraternal dizygotic twins (who are more likely to share environmental risk factors) and 50% for identical monozygotic twins who share identical genetic backgrounds. Reproduced with permission from WH Freeman.[6]

It is interesting that the first-degree relatives and fraternal twins both share on average 50% of their genes, but the non-twins (ie brothers and sisters of different ages) have a lower risk of being concordant for schizophrenia, possibly because of a greater divergence in their environmental circumstances. The principle of comparing monozygotic and dizygotic twins (ie siblings of same age, sharing 100% and 50% genetic material respectively) is widely used by geneticists as they try to estimate the proportion of a complex trait accounted for by genetic influences. In practice, a measure called 'heritability' is calculated representing the proportion of the variability (strictly the total variance, V_T) within the trait that is due to genes. It depends somewhat on the circumstances under which it is investigated, but it is a very powerful approach.

The final outcome of all this is that for any trait that, like behaviour, has many factors underpinning it, including the contribution of numerous genes and environmental influences, the characteristics determining the appearance of the trait (phenotype) are going to be distributed normally (see Figure 3.3). Apart from one or two confounding factors, this is generally true for the majority of characteristics, including height, weight, intelligence etc. In addition, there is also a phenomenon known as the 'threshold' effect, whereby the underpinning factors may add up in a particular pattern, so that it is only when we get to an unusual combination of these that we have a manifestation of the trait in question. This is thought to apply to many behavioural phenotypes, including schizophrenia and aggression. Generally speaking, we now consider many such conditions to exist at the extreme of a continuum with normal behaviour. As a result of developments both in our understanding of how to evaluate the genetic contribution to behaviour and in the Human Genome Project, the two research areas have now converged in an attempt to identify the individual genes contributing to the overall picture. These are genes referred to as quantitative trait

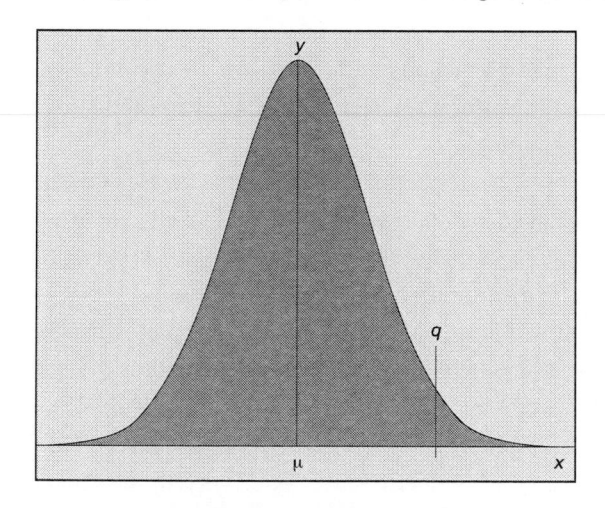

Figure 3.3 Illustration of a normal distribution of a behavioural trait. The characteristic bell-shaped curve is typically found for the distribution of many quantitative traits, including some behavioural phenotypes. In such cases the y axis is a scale representing the number of individuals with a given trait value and the x axis the trait value itself. Some clinically defined behaviours, however, are thought to have a continuous distribution of underlying predisposing factors (both environmental and genetic). In this case the x axis can be thought of as representing a scale of predisposing factors. Only when their combined effect reaches a particular threshold does the phenotype in question arise. μ lies on the mean of the distribution. q denotes the threshold beyond which manifestation of the trait is considered to be a psychological disorder.

loci, or QTLs. In practice, this means that the behaviour geneticists are now looking both for the individual genetic factors that provide an input into the distribution at the quantitative level and also at how these interact with one another and with the environment. QTL investigation is currently, therefore, at the forefront of behavioural genetic research.

Returning to an earlier point, if moral values have a genetic basis, then the moral values must, in some sense, either conform to what we would expect of them after undergoing evolution through natural selection or have arisen by a random accumulation of genetic factors which has then determined moral behaviour. The latter seems less likely and we should therefore attempt to examine how geneticists might view the concept of an evolutionary origin for the development of morals. In Kohlberg's stages 1 and 2, there appears to be the early emergence of moral behaviour, but one that is purely self-interested because it aims to avoid punishment and attain reward. It is mindful of the immediate consequences of actions and of the fact that one's needs or interests may sometimes best be served if one recognises that other people have their needs and interests too. An individual at this stage is aware that he needs others' services and/or good will for his own ends.

Kohlberg's stage 2 is interesting in evolutionary as well as cognitive terms. In essence, the tit-for-tat strategy of cooperation embodies the 'you scratch my back and I'll scratch yours' mentality. Now, this is really the vernacular for what is known as 'reciprocal altruism'. It is explained evolutionarily by the fact that if you do a favour for somebody, it is because you rely on the probability that somebody, at some later point, is going to do a favour for you. Therefore, you need not be related to them and it is not necessary to invoke the traditional explanation of altruistic behaviour, which is one of 'kin selection'. In the case of human society, altruistic behaviour confers an evolutionary advantage because you are likely to bump into the same people on a repeated basis and have many interactions with them. For this reason, it is going to work better for you to perform the favour, with the expectation of a favour being returned in the future.

Kohlberg's stages 3 and 4 see the development of an aim to please and gain approval from others as well as to maintain a relationship of loyalty and trust. This aim, along with the awareness that others have conflicting interests, presupposes the sensory development of a concept of theory of mind – that is, the ability for an individual within a relationship to take on board what the other person is thinking. In normal children, this 'theory of mind' is thought to develop between the ages of 3 and 4 years old, and yet if we look at disorders such as autism, for example, one of its key features is a failure to build up this theory.[7] Autism has a very high heritability with a very strong genetic flavour. This provides clear indirect evidence that genetic factors have some role to play in the development of this stage of morality.

Stages 3 and 4 introduce concepts of the laws and norms of a larger social system beyond simple reciprocal altruism. A slightly more sophisticated evolutionary model therefore becomes necessary to accommodate the additional complexities. The model is that of an individual adopting an evolving pragmatic strategy, which may be overt or subconscious. The question is: if you come across people who you can be virtually sure you will never meet again is there any evolutionary advantage to behaving morally towards them, even if it is at some cost to yourself? There are two main factors that suggest that it *would* be to your advantage and that seem to fit in with the concepts of both a moral framework and evolutionary advantage. The first of these concepts is loss of reputation: if you do not behave well, you will lose your good reputation and that may affect the way people, even those not directly involved in a particular moral transaction, are going to treat you. The second concept is fear of punishment.

The idea of reputation has been proposed by several people over the years to explain how cooperation in large groups can emerge, where there are many interactions, but no guarantee of meeting the same individuals repeatedly. It allows individuals to be identified as 'cooperators', so that others can recognise them as 'good' people to interact with. There is therefore a reason to treat them well on any encounter. A compelling account of this type of scenario was published in *Nature* by Milinski and colleagues.[8] They set up a very interesting series of exchanges based on conventional game theory. Participants were asked to play a game in which there is a conflict between the self and the common good, whereby if individuals cooperate with others, the group will survive very successfully, but if one individual cheats then he may do better than the rest but compromise the group. This is the 'tragedy of the commons' concept.[9] For long-term success and to ensure that the pasture will remain in good condition, the commons should be preserved through cooperation. Nevertheless, in the short term it makes sense for a herder to allow his animals to graze it as fast as possible, without worrying about the other herders; sometimes it is better in the short term to cheat than to work for the common good.

Milinski's experimental game contained a common good element, but individuals could benefit by cheating and going their own way. Usually when such games are played, over a period of time the participants' cooperation drops dramatically. In Milinski's experiment, however, the game was alternated with another in which a reputation could be established. So, if an individual was seen to be responsible and cooperative, people recognised them as 'good' and they established a reputation as such. It was found that by alternating the 'public goods' game and the game enabling individuals to gain a reputation, cooperation was maintained in the community at a very high level. Further, if individuals cooperated with someone in a preceding round of public goods, then the probability of receiving a refusal in the next round was much lower than if they had refrained from cooperating, emphasising the value of reputation in an 'open' society. Interestingly, however, when subjects were told that they did not have to alternate between the two and could participate only in the public goods game, the cooperative system crashed very rapidly.

In evolutionary terms, there is an overall benefit to individuals if reputations can be established, and a high level of morality can be maintained in society. This, however, is dependent on the members of the population being involved in many interactions. There are going to be some circumstances in which only very few opportunities exist for interaction, and these may be thought to provide little chance for building a reputation. Even in these cases, though, there may be an impetus to behave morally if the concept of punishment comes into operation. Punishment is, of course, one of the mechanisms that societies have adopted to maintain moral stability. (Fehr and Gachter drew attention to this aspect in a paper entitled 'Altruistic punishment in humans'.[10]) One of the difficulties associated with the concept of using punishment as a control mechanism, however, is that somebody has to accept the burden (or cost) of undertaking the punishing. How can this enable a stable situation to arise?

In the study by Fehr and Gachter,[10] a game was organised in which participants played with money and cooperation was encouraged for success. As is common in such game constructs, the system could also be exploited by an individual through a limited number of 'cheats'. But there was a twist: as well as knowing who had defaulted, individuals could also punish the defaulters – effectively by fining them. The punishers lost money by undertaking the punishing, but not as much as the cheats. The study found that people who deviated from an established mean cooperation value got punished a great deal. Participants were therefore afraid to default, which led to a fairly stable situation. These conclusions are rather simplistic, but they have to be because we are dealing with game theories and not real life. Nevertheless, it seems that many types of behaviour which can be labelled as moralistic can also be explained through direct or indirect *self*-interest survival models. Therefore, it can be posited that they should fall within the ambit of natural selection and have evolutionary relevance.

Another problem that such models have to overcome is that if an individual is particularly clever, they can avoid all the checks to good behaviour and become a sophisticated cheat. If you think about all real societies, though, they often do have a certain number of sophisticated cheats. It is up to the majority of conformers to decide what are the acceptable categories and numbers of people allowed to contravene the normal moral boundaries. In practice, it may be necessary to accept that the system has to cope with a certain amount of 'noise' in terms of morality levels.

Molecular genetics, morality and relevant behaviours

The ideas discussed above might be termed 'soft science' approaches, but there might also be evidence for a genetic programme involved in behaviours, including moral behaviour. Advances in molecular and statistical genetics have paved the way for many groups to begin the search for genes involved in complex traits, including intelligence and emotionality, and it is possible that some aspects of these characteristics have direct relevance to moral behaviour. The final section of this chapter will examine the evidence supporting this contention.

Genes for behaviour do exist: it has been established that single genes can control disorders with profound behavioural disturbance which may compromise accepted moral standards such as in Huntington's disease; indeed the gene responsible has in this case been isolated and characterised. As yet, however, no single genes have been isolated which have a major impact on, for example, socialisation, intelligence or civility. As we saw earlier, there is a very good reason for that, since the majority of these and similar traits are very sophisticated, undoubtedly complex and involve many different genes interacting with each other and with the environment. It takes only one of the genes which underpin a given behaviour to go wrong through mutation for the whole system to break down. Therefore it is relatively easy to find genes that completely block normal functioning, but to find genes that slightly enhance or impair an aspect of that normal functioning, such as intelligence or morality, is very difficult. In other words, 'Any jackass can kick down a barn, but it takes a carpenter to build one.'[4]

So how has the human genome programme contributed to our understanding of the genetics of morality? There are approximately two metres of DNA in the nucleus of every cell, and these two metres represent up to three billion base pairs, which have to be identified and placed in order to give an entire account of the human genome. Figure 3.4 illustrates how rapidly molecular genetic technology has developed, starting in 1953 with the characterisation of the structure of DNA and its pivotal role in hereditary mechanisms. Rapid DNA sequencing was, however, possible only following

2004	The rat genome is sequenced
2002	The mouse genome is sequenced
2001	Publication of the human genome sequence
2000	Human genome sequenced and assembled
1999	The *Drosophila* genome is sequenced
1998	The worm *C. elegans* is sequenced
1996	An archaeon (and extremophile) is sequenced
1996	The yeast genome is sequenced
1995	A free-living organism, *Haemophilus influenzae*, is sequenced
1991	J Craig Venter describes a fast new approach to gene discovery using expressed sequenced tags
1986	Leroy Hood develops the automated sequencer
1986	Launching the effort to sequence the human genome
1983	Kary Mullis conceives and helps develop polymerase chain reaction
1978	David Botstein initiates the use of restriction fragment length polymorphisms
1977	Walter Gilbert and Frederick Sanger devise techniques for sequencing DNA
1973	Herbert Boyer and Stanley N Cohen develop recombinant DNA technology
1972	Paul Berg creates the first recombinant DNA molecules
1970	Hamilton O Smith discovers the first site-specific restriction enzyme
1970	Howard Temin and David Baltimore independently discover reverse transcriptase
1969	Jonathan Beckwith isolates a bacterial gene
1967	Mary Weiss and Howard Green employ somatic cell hybridisation to advance human gene mapping
1961	François Jacob and Jacques Monod develop a theory of genetic regulatory mechanisms
1961	Marshall Nirenbera cracks the genetic code
1957	Francis HC Crick sets out the agenda for molecular biology
1956	Arthur Kornberg crystallises DNA polymerase, the enzyme required for synthesising DNA
1953	Francis HC Crick and James D Watson discover that the chemical structure of DNA meets the unique requirements for a substance that encodes genetic information

Figure 3.4 History of molecular genetics timeline. The chronology of important events in the development of contemporary molecular genetics is shown. The success of the Human Genome Project has been critically dependent on a number of major technological developments; arguably, however, the sequencing reaction devised by Frederick Sanger has been pivotal. Reproduced with permission from J Craig Venter Institute (www.venterinstitute.org).

the publication of the so-called di-deoxy approach by Fred Sanger, and of the 'chemical cleavage' strategies by Walter Gilbert, both in 1977.[11,12] Fred Sanger is one of the very few people to have been awarded two Nobel Prizes. The first was for working out how to sequence a protein, and the second for DNA sequencing technology. His technique has continued to be used, leading to the publication of the draft of the human genome sequence in 2001 and of the mouse genome in 2002.[13] The sequences of a variety of simpler genomes have been completed along the way, including the bacterial genome in 1995 and the fruit-fly, *Drosophila melanogaster,* in 2000.[14]

Until fairly recently, it was generally thought that, as befits a pretty complicated and evolutionarily sophisticated organism, the number of genes that each human has may be somewhere around the 100,000 mark. The sequencing of the human genome, however, now indicates that the number is considerably smaller – somewhere between 30,000 and 40,000. Some people found this figure a big disappointment, thinking that we should have more genes, and certainly more than, for example, the apparently simple fruit-fly (which has around 15,000). It is worth remembering, though, that a similar number of genes in two organisms does not necessarily mean that they have a similar level of complexity. This is because most of our genes are made up of little sections, which are not continuous in the human genome. When the sections are turned into the messages, those sections can be put together in different combinations by a process known as differential splicing. On average there are probably about three times as many different gene products (*viz* proteins) that can be produced as there are individual genes.

Another important observation to come from sequence analysis of the human genome is that about one base in every thousand may be different in different individuals. This means that we have more than 6 million sites along our chromosomes that have the potential to vary amongst the population. Some of this variation may be highly significant – particularly if it occurs in the coding sequence of genes. Some of it, occurring elsewhere, may not be functionally significant but can provide useful landmarks for mapping the genome. Overall, it seems that most genes have variant forms at significant levels in the population and that this heterogeneity is likely to be represented in the QTLs referred to earlier. Any complex character, therefore, probably arises from the combination of a multiplicity of potentially variable sites at a genetic level.

What, then, is the significance of this in the context of discovering genes for behaviour and, possibly, morals? We can take two basic approaches. The first is to make use of our new-found knowledge of the genome to make educated guesses as to what kind of genes are *likely* to be important in causing behaviour. This is referred to as a 'candidate gene' approach. In the Social, Genetic and Developmental Psychiatry (SGDP) Centre in London, we have elected to concentrate on sets of genes involved in neurotransmitter pathways. So, instead of just looking at genes one at a time, we are looking at all the genes involved in integrated pathways of neurotransmission by examining genes controlling the steps involved in the metabolism, turnover and reception of neurotransmitters at the nerve synapses. Our goal is to investigate how variation in these genes affects behaviour.

This candidate gene approach has had some limited success and it could be applied in the context of morality. Instead of looking for a gene for morality *per se*, one could investigate types of behaviour and personality that contribute to moral character, such as aggression, which is to some extent at least moderated by genes. The work led by Temi Moffit and Avshalom Caspi (also at the SGDP Research Centre) illustrates this point well. Their team has been investigating a cohort of individuals from Dunedin, New Zealand, who have been examined in great detail every three years or so since their birth. From these and other studies, it is clear that behaviours which are characterised by deviation from commonly accepted moral norms in terms of social behaviour in childhood and adolescence may lead on to criminality and violent aggression in adults. Such problems in childhood can also overlap with a condition known as attention deficit and hyperactivity disorder (ADHD), which is widely recognised to have some genetic aetiology. So it is possible that genes can predispose young children to difficult, inattentive behaviour, and this in turn may progress to antisocial behaviour later in life.

It has been suspected for some time that the dopamine pathway is quite central to some aspects of ADHD because Ritalin (methylphenidate hydrochloride), a drug used to treat ADHD, is successful in a proportion of patients, and appears to act by blocking the re-uptake of dopamine at the synapses connecting nerves together in the brain. One of the important genes in the dopamine pathway is the DRD4 dopamine receptor, and work reviewed recently by Steve Faraone *et al* (2001) has shown that one of the versions of the gene DRD4 appears to predispose individuals to ADHD.[15] This is one of the few consistently replicated observations of individual genes being implicated in the aetiology of behavioural disorders.

In an interesting development on this topic, it has been reported that the mutation that caused this variant form of DRD4 occurred only about 40,000 years ago, not very long in human population terms. Yet this variant is now quite widespread – more so than one would expect by chance. One explanation might be that hyperactivity in a general sense has been selected, so individuals who were a little bit aggressive, or a little bit hyperactive, might have been at an advantage in the evolving human population and the novel form of DRD4 would have been selectively propagated.[16]

Another gene we have been looking at, which again is involved in neurotransmitter metabolism, is monoamine oxidase A (MAOA), which metabolises both dopamine and serotonin. The MAOA gene is on the X chromosome, and so a male has no choice as to which version is expressed. Females, on the other hand, have got two copies, and one 'good version' can over-ride the effects of a defective version. There is a much-referenced paper dealing with this enzyme.[17] It describes a Dutch family pedigree, with a complex behaviour pattern involving some, or all, of the following features: mild mental retardation, a tendency to commit arson, aggression and inappropriate sexual attention to family members. This pedigree (see Figure 3.5) shows that boys who are affected clearly segregate in the expected manner for an X-linked recessive gene. Further examination demonstrated that every one of the affected boys carries a (nonsense) mutation that completely stops their MAOA gene from working. These boys therefore have no monoamine oxidase A activity. 'Carrier females' are not affected because they have another X chromosome, which can express a 'good' version of the gene, covering for the deficiency.

As well as this extremely rare condition that Brunner found, where the gene is completely inactive, we have been working recently on behaviours that appear to be influenced by variant forms of the MAOA gene. These forms are found at high frequency in the normal population. We, and others, have shown that that there are versions of the MAOA gene which can be classified as high- and low-activity forms. In a study, again coordinated by Professors Caspi and Moffit at the SGDP Centre, we have been looking at the influence of the MAOA genotypes on patterns of behaviour. We have been able to

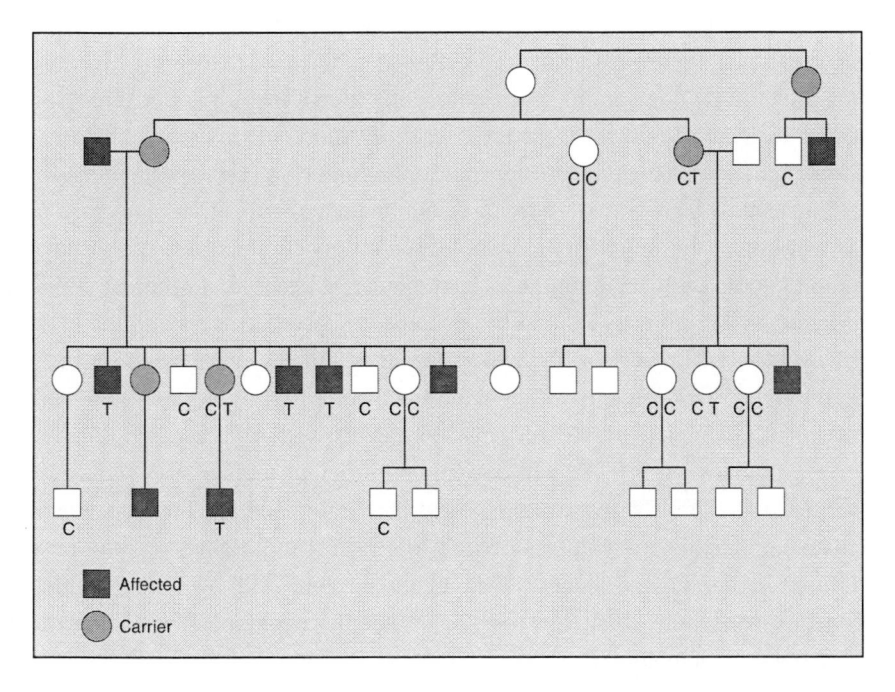

Figure 3.5 A pedigree showing the inheritance pattern of an X-linked recessive disorder associated with antisocial behaviour. Brunner et al (1997) described a Dutch family in which affected males exhibited complex antisocial behaviour and who had a defect in their monoamine oxidase gene.[16] This is carried by the X chromosome and, like haemophilia, the resulting deficiency is expressed in males and transmitted by unaffected, carrier females. C and T refer to the defining base changes. The presence of T (thymine) rather than C (cytosine) results in a totally defective protein. Reproduced with permission from the Novartis Foundation.[18]

analyse DNA samples from the Dunedin cohort of individuals and this has led to the discovery that males with the low-activity form of MAOA are more likely to develop antisocial behaviour following abusive childhoods.[19] The high-activity version, in contrast, appears to confer protection against such disadvantageous environments. To our knowledge, this is the first time that a specific interaction between a single genetic factor and an environmental variable, predisposing individuals to a particular behaviour, has been adequately documented.

Apart from examining genes that encode proteins with properties which make them sensible candidates to be involved in behaviour, the other way to find genetic factors implicated in a behaviour is to start without *a priori* assumptions and search for them throughout the whole genome. Professors Plomin (SGDP) and Owen (Cardiff) have coordinated studies that have sifted through the genome for any chromosomal regions that affect cognition. General cognition, or 'g', represents the common factor involved in a range of skills such as language and mathematical ability, and which again has obvious relevance to the practice of morality. In the study, a large number of people of average intelligence are compared with similar numbers of those who have IQs about two standard deviations above average. Any positive findings are then explored in more detail and attempts made to replicate the findings with additional sample groups of people with average and very high intelligence (more than three standard deviations above average). At the practical level, this is achieved by probing the genome in about 2,000 different places to pick up any differences in chromosomal sites between the high and the medium IQ populations. Each apparently positive result is put through a series of increasingly stringent tests for replication. At this stage, there are some markers for regions of DNA that look interesting. Before becoming too enthusiastic, however, we are trying to eliminate any errors that may creep in as a result of the populations having different genetic features, which have nothing to do with cognition but which may arise because of different ethnic origins. As yet, no markers for chromosomal regions have survived the most stringent tests. However, only 2,000 sites along the chromosomes have been examined so far. Contemporary evidence suggests that in order to achieve a reasonably complete survey for cognition genes, somewhere around 50,000 or maybe between 500,000 and one million different marker sites should be surveyed – a project that would be extremely demanding, but one which advancing technology will bring within the realms of possibility in the near future. Buther *et al*[20] have indeed identified four SNP variations which together contribute about 1% to the variance in cognition, by scanning the genome with 10,000 markers.

The interesting point to arise from this is that, in principle, and if one could collect and compare groups of individuals from either end of the spectrum of perceived moral attributes, one could use this genome-scanning approach to search for genes for morality.

One other interesting story, which may be of some significance in considering 'moral' behaviours from an evolutionary perspective, is to do with a phenomenon known as imprinting. It seems that, over evolutionary time, it has proved advantageous to switch off perfectly intact copies of particular genes when they are passed through a female line during gamete production, whilst other genes are switched off when passed through the male line. The emergence of this mechanism of 'imprinting' has been attributed in at least some instances to the competition between the male and female for genetic influence on the child's growth rate. Males want their offspring to be large so that they compete successfully. Mothers also want their progeny to be reasonably sized, but only to an extent that is not going to prevent them from having further children. There are genes which make the fetus grow rapidly and which remain capable of activity when transmitted by the male but are switched off by the mother. This keeps the growth of the fetus within acceptable limits and prevents all her 'resources' being taken by the child of a single father, which is important given that, at the stage in evolution when these mechanisms evolved, the mother's next child would probably have been by a different father. So there appears to be a kind of 'sex war' associated with imprinting. To illustrate, imagine we are considering maternal imprinting so that when the mother produces gametes, the locus in question (marked A1 in Figure 3.6) is switched off, whereas in the male it is not. The offspring will show one genotype or the other and will not be able to express the product encoded by the female genotype at all. The genes that are being switched off are perfectly functional normal genes and, because of their anomalous expression patterns, they cause significant changes to the anticipated inheritance pattern.

David Skuse's work at the Institute of Child Health illustrates how the phenomenon of imprinting has implications for our understanding of behaviours that have a moral dimension.[21] He has investigated a series of girls with Turner's syndrome who have only a single X chromosome and lack the usual second chromosome. If the X chromosome in a girl with Turner's syndrome comes from the father, the rarer situation, the girl tends to have much better social skills and sensitivity to feelings of others (as measured, for example, by how good she is at interpreting photographs of faces registering a variety of emotions). Those girls who get the X chromosome from their mother, on the other hand, perform much more poorly on these tasks. What this may suggest is that, as a result of maternal imprinting, a gene (or genes) encoding aspects of social skills is on the

Figure 3.6 Schematic representation of maternal imprinting. This follows a gene (designated A) located on a homologous pair of chromosomes. A female in whom imprinting occurs to the gene during gamete formation is crossed with a male who is heterozygous for a normal copy of gene A (A1) and a defective version (A2). The offspring of such a coupling may inherit one good version of the gene from the father or one defective version. As imprinting inactivates the maternal version of A, half of the offspring will have no functional copy of the gene and thus be affected by a disorder.

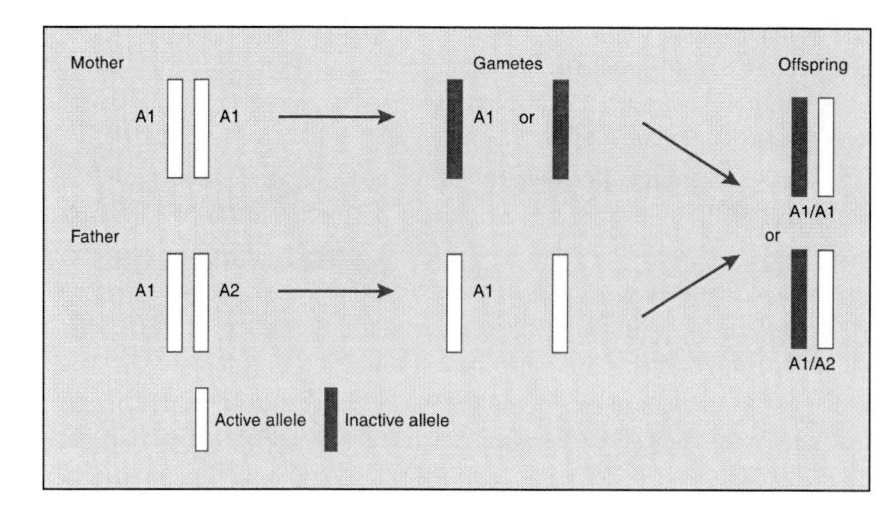

X chromosome and is switched off during the formation of female gametes but is transmitted unimpaired during male gametogenesis. The implications of this are rather interesting, given that normal boys receive their only X chromosome from their mother. Boys, then, have only this imprinted (switched off) version of the social skills gene or genes and would therefore be expected to behave differently from girls for this particular behavioural trait. This intriguing model has yet to be validated at the molecular level, however.

Finally, there is a powerful way to look at behaviours by using animal models and performing gene 'knock-out' experiments in mice. This procedure enables the experimenter to remove particular genes, and to study any resulting changes in the behaviour patterns of the mice. As a result of that sort of experimental protocol, a series of imprinted genes has been found, and two of the most interesting are known as *Peg* I and *Peg* III.[22,23] They are maternally imprinted (inactivated) so that if the progeny also get a null allele from the father (achieved by gene knock-out in the father) they are unable to make any gene product. In such cases, the female offspring show very abnormal behaviour, frequently failing to build nests and abandoning the pups. In human terms, of course, a mother who neglects her children would generally be perceived as having low moral standards. It is therefore remarkable that apparently complex behaviour with moral overtones can be so profoundly affected by a single gene.

As we can see, then, most types of moral behaviour can be modelled from an evolutionary perspective in a very simplistic way as being adaptive for survival in communities where cooperation, reputation and punishment must all be taken into consideration. There is, in addition, indirect evidence supporting a very strong genetic input into the behaviours and personality types that affect whether a person is seen as 'moral'. There are of course people who object to the reduction of human morality to an evolutionary mechanism that is centred, at its core, on self-interest. The concept of the human conscience, and the fact that one experiences emotions of guilt, sympathy and so on first hand, make it difficult to accept that altruistic actions are in essence nothing more than a strategy by which a person maintains the best possible situation for himself in a society requiring cooperation. Yet, it is still possible that these subjective emotions are by-products of evolutionary development, and one could propose that morality, together with its attendant emotional responses, can be viewed as a complex trait similar to intelligence. Morality may therefore be amenable to similar research strategies that attempt to unravel the contribution made by genes as well as the impact of upbringing. This may be the starting point on a very long road to finding genes that have a significant impact on moral behaviour.

References

1. Piaget J. *The moral judgement of the child*. New York: The Free Press, 1965.
2. Kohlberg L. Stage and sequence: the cognitive-developmental approach to socialisation. In Goslin DA (ed) *Handbook of socialisation theory and research*. New York: Rand McNally, 1969.
3. Turiel E. *The development of social knowledge: morality and convention*. New York: Cambridge University Press, 1983.
4. Pinker S. *How the mind works*. London: Penguin Press Science, 1997.
5. McGuffin P, Huckle P. Simulation of Mendelism revisited: the recessive gene for attending medical school. *Am J Hum Genet* 1990;46994-999.
6. Gottesman I. *Schizophrenia genesis: the origins of madness*. WH Freeman: New York, 1991.
7. Baron-Cohen S, Leslie AM, Frith U. Does the autistic child have a 'theory of mind'? *Cognition* 1985;21:37-46.
8. Milinski M, Semmann D, Krambeck HJ. Reputation helps solve the 'tragedy of the commons'. *Nature* 2002;415:424-6.
9. Hardin G. The tragedy of the commons. *Science* 1968;162:1243-8.
10. Fehr E, Gachter S. Altruistic punishment in humans. *Nature* 2002;415:137-40.
11. Sanger F, Nicklen S, Coulson AR. DNA sequencing with chain-terminating inhibitors. *Proc Natl Acad Sci U S A* 1977;74:5463-7.
12. Maxam AM, Gilbert W. A new method for sequencing DNA. *Proc Natl Acad Sci U S A* 1977;74:560-64.
13. Waterston RH, Lindblad-Toh K, Birney E *et al*. Initial sequencing and comparative analysis of the mouse genome. *Nature* 2002;420:520-62.
14. Adams MD, Celniker SE, Holt RA *et al*. The genome sequence of *Drosophila melanogaster*. *Science* 2000;2185-95.

15 Faraone SV, Doyle AE, Mick E, Biederman J. Meta-analysis of the association between the 7-repeat allele of the dopamine D(4) receptor gene and attention deficit hyperactivity disorder. *Am J Psychiatry* 2001;158:1052–7.

16 Ding YC, Chi HC, Grady DL *et al.* Evidence of positive selection acting at the human dopamine receptor D4 gene locus. *Proc Natl Acad Sci U S A* 2002;99:309–14.

17 Brunner HG. MAOA deficiency and abnormal behaviour: perspectives on an association. *Ciba Found Symp* 1996;194:155–64.

18 Bock G and Goode JA (eds). Genetics of criminal and antisocial behaviour. Chichester: John Wiley & Sons Ltd, 1996.

19 Caspi A, McClay J, Moffitt TE *et al.* Role of genotype in the cycle of violence in maltreated children. *Science* 2002;297:851–4.

20 Butcher LM, Meaburn E, Knight J *et al.* SNPs, microarrays and pooled DNA: identification of four loci associated with mild mental impairment in a sample of 6,000 children. *Human Molecular Genetics* advance access, March 2005.

21 Skuse DH, James RS, Bishop DV *et al.* Evidence from Turner's syndrome of an imprinted X-linked locus affecting cognitive function. *Nature* 1997;387:652–3.

22 Lefebvre L, Viville S, Barton SC, Ishino F, Surani MA. Genomic structure and parent-of-origin-specific methylation of Peg1. *Hum Mol Genet* 1997;6:1907–15.

23 Lefebvre L, Viville S, Barton SC *et al.* Abnormal maternal behaviour and growth retardation associated with loss of the imprinted gene Mest. *Nat Genet* 1998;20:163–9.

4

The deceiving brain

SEAN A SPENCE

Professor Spence proposes that the default state of the human mind is truth telling. His team examined the response times in individuals instructed to lie or tell the truth and carried out functional brain scanning during the same exercise. They were able to identify a 'lie centre' in the ventrolateral prefrontal cortex of the brain indicating a possible mechanism for moral behaviour. Professor Spence discusses the survival advantages of deception.

Introduction

Why might deception interest a clinical readership? Is it not a moral issue, more relevant to legal or theological discourse? Might we ever be able to scientifically determine whether another human being is lying to us? And should we want to anyway?

The answers to these questions may vary considerably depending upon the envisaged clinical setting, but in psychiatry, neurology, medicolegal practice and perhaps other areas of medicine, doctors are sometimes called upon to judge the veracity of their patient's account (even if this is not always made explicit). Consider the distinction between feigned or malingered physical symptoms and those ascribed to hysteria or conversion disorder: what objective grounds are there for differentiating between these diagnoses?[1] It would seem that there is little objective evidence that would favour one over the other and the diagnosis made may all too easily depend upon the physician's opinion of the patient's personality or social background. Also, the subtle 'tricks' used to elicit hysterical motor inconsistency (eg movement of the 'paralysed' limb) might just as well be diagnostic of deception.[2] When making these judgements, the physician is often being called upon to guess the patient's intentions.

The prospects of success for such an enterprise appear slim. In psychological studies, subjects attempting to judge whether others are lying to them often perform at the level of chance.[3,4] With the possible exception of secret service agents, it seems to make little difference whether the putative 'lie detector' is a judge, a police officer or a doctor as they are equally inaccurate. With regard to secret service agents, it may be that those engaged in day-to-day front-line witness interrogation perform better than chance whereas their superiors, exercising more managerial duties, perform less well.[3] Lie detection may therefore be a skill requiring frequent rehearsal.

There is a more subtle aspect to the study of lying, however, which might elude those unaccustomed to conceptualising human behaviour in terms of its higher cognitive control processes. For it would appear that deception behaves as if it is a skill (akin to lie detection) which must be worked at, which requires attention and where fatigue may lead to inconsistency or unintended confessions.[4]

It is significant that lying engages higher brain systems because such systems may be differentially affected in neuropsychiatric diseases and in some of the most difficult patients whom psychiatrists see, for example psychopaths and sex offenders, there is scant evidence of intellectual impairment. Predicting the future is not a solely intellectual exercise, sometimes the actions of a 'good' deceiver may have dire consequences for others.

This chapter examines whether deliberate deception comprises an 'executive' task, primarily engaging higher brain systems, and whether brain imaging might ever tell us what someone else is thinking.

Learning to lie

[L]ie... a false statement made with the intention of deceiving...[5]

[D]eception... a successful or unsuccessful deliberate attempt, without forewarning, to create in another a belief which the communicator considers to be untrue.[4]

Judging by the instructions offered by religious texts dating from antiquity, lying and deception have been of concern to humans for millennia.[6] However, despite the apparent premium placed upon honesty in ancient and modern life, there are emerging literatures in evolutionary studies,[7] child development and developmental psychopathology suggesting that the ability to deceive is an acquired and normal ability. Such behaviours follow a predictable developmental trajectory in human infants[8] and are impaired in those with specific neurodevelopmental disorders such as autism.[9] As a result there would appear to be an interesting tension between what is socially undesirable but 'normal' (ie lying) and that which is socially commendable but pathological (eg always telling the truth). Higher organisms have evolved the ability to deceive each other,[10] while humans in a social context are openly encouraged to refrain from deception. Of course, it might be hypothesised that it is precisely because the human organism has such an ability to deceive that it is called upon to exercise control over its potential use.

What's the use of lying?

Given the 'normal' appearance of lying and deception during childhood, a number of authors have speculated about the purpose served by such behaviours. One view has been that deceit delineates a boundary between self and other, originally between the child and the mother.[11] Knowing something that the mother does not know establishes for the child the limit of the former's omniscience, ceding relative power to the latter. Indeed, such power over information might drive the pathological lying seen among dysfunctional adolescents and adults.[11]

Lying also eases social interaction, by way of compliments and information management. Truthful communication at all times would be difficult and perhaps rather brutal.[4] It is therefore unsurprising that 'normal' subjects admit to lying most days.[4] Social psychological studies, often of college students, suggest that lying facilitates impression management, especially early on in a romantic relationship, although parents continue to be lied to frequently.[4]

Deception is a vital skill in the context of conflict especially between social groups, countries or intelligence agencies. When practised under these circumstances it might also be perceived as a 'good'. When one is branded a liar, however, the advantage formerly gained may be lost. Although fluent liars might, at times, make entertaining companions, being known as a liar is unlikely to be advantageous in the long term.[4]

Principles of executive control

Control of voluntary behaviour in everyday human life is crucial but likely to be constrained by cognitive neurobiological resources.[12] Control (or executive) functions are not necessarily 'conscious', although they may access awareness.[13,14] Executive functions include planning, problem solving, the initiation and inhibition of behaviours and the retention of useful data in conscious working memory (eg the telephone number about to be dialled). These functions engage specific regions of prefrontal cortex (PFC), although they are also distributed across several brain systems. There seems to be, however, a principle to the cognitive architecture of executive control: higher centres, such as PFC, are essential in adaptive behaviour in novel or difficult circumstances, while lower, slave systems, implicating posterior or sub-cortical systems, may be sufficient to perform routine or automated tasks (eg riding a bike while thinking of something else).[15]

A recurring theme in the psychology of deception is the difficulty of deceiving in high-risk situations: information previously divulged must be remembered, emotions and behaviours controlled and information managed.[4] The latter are quintessentially executive functions. Much of the behaviour of the liar may, therefore, be seen from a cognitive neurobiological perspective as located on a continuum with other situations where behavioural control is exerted, albeit using limited resources. Some clinical examples may serve to illustrate the principles underlying such behavioural control.

Control in the clinic

One of the means by which a psychiatrist may assess whether a patient with schizophrenia who is receiving neuroleptic medication exhibits involuntary movements is through the use of distraction, eg when the patient is requested to stand and perform complex hand movements. While distracted by the manual task the patient may begin to shuffle repetitively on the spot and his tongue may begin to protrude, exhibiting dyskinesia.[16] While executive, control systems are engaged in complex tasks they fail to inhibit other, involuntary movements (eg of the legs and tongue). Similarly, a patient with hysterical conversion symptoms, such as motor paralysis, may move the affected limb when they are distracted or sedated, suggesting that executive processes were engaged in maintaining the functional symptom when it was present.[1]

In certain situations, liars may also reveal their deception by their bodily movements. While telling complex lies, for example, they may make fewer hand and arm movements, otherwise known as 'illustrators'.[4] Slower, more rigid behaviour exhibited by liars has been termed the 'motivational impairment effect', and police officers have been advised to observe witnesses from head to toe rather than focusing upon their eyes.[4] Indeed, it is tempting to reconsider the advice of one French detective in this context:

> *Commissairre Guillaume, like Maigret, had a dislike of violent interrogations and told* [George] *Simenon that he had found a more effective method for eliciting the truth. He would make a suspect strip naked in front of a room of fully clothed detectives. 'They don't tell lies for long in that costume,' said Guillaume.*[17]

No doubt Guillaume's method was not concerned primarily with eliciting the motivational impairment effect.

Lying as a cognitive process

Deceiving another human subject is likely to involve multiple cognitive processes, including social cognitions concerning the victim's thoughts (their current beliefs) and the monitoring of responses made by both the liar and the victim in the context of their interaction. In light of what has been described above, it is possible to posit that the liar is called upon to do at least two things simultaneously: to construct a new item of information (the lie) and to withhold a factual item (the truth, assuming that the liar knows and understands what constitutes the 'correct' information). Within such a theoretical framework it is apparent that the truthful response comprises a form of baseline, or pre-potent response: that which we would expect to be provided if an honest subject were asked the question, or the liar was distracted or fatigued. We might, therefore, hypothesise that responding with a lie demands something extra, that it will engage executive prefrontal systems more so than telling the truth. Hence, we have a hypothesis that we can test using functional neuroimaging.[18]

Imaging deception

In our own work we hypothesised that the generation of 'lie' responses (in contrast to 'truths') would be associated with greater prefrontal activity,[18] and that the concomitant inhibition of relatively pre-potent responses ('truths') would be associated with greater activation of ventral prefrontal regions (systems already known to be implicated in response inhibition).[19]

We used a simple computerised protocol in which subjects answered questions with a 'yes' or a 'no', pressing specified single computer keys.[18] All the questions concerned activities that the subjects might have performed on the day they were studied. When they were first interviewed we had acquired information about that day's activities. The subjects then performed these tests in the presence of an investigator, who was required afterwards to judge whether the subjects' responses were truths or lies. The computer screen presenting the questions to the subjects also carried a green or red prompt (the sequence counterbalanced across subjects). Without the investigator knowing the colour rule, subjects responded with truth responses in the presence of one colour and lie responses in the presence of the other. All questions were presented twice, once each under each colour condition, so that in the end we were able to compare response times and brain activity during different responses. We studied three cohorts of subjects in the laboratory (30–40 subjects in each)[6] and one sample of 10 subjects in the MR,[18] performing two variants of the experimental protocol (establishing internal validity).

The brain imaging technique applied was functional magnetic resonance imaging (fMRI).

Our analyses revealed that whether samples were studied inside or outside the scanner there was a statistically significant effect of lying upon response time (approximately 200 ms longer compared with responding truthfully). In the scanned sample, there were foci of increased activation in bilateral ventrolateral prefrontal and anterior cingulate cortices associated with 'lie' responding.[18] The data support the hypothesis that prefrontal systems exhibit greater activation when subjects are called upon to generate experimental 'lies' and that longer processing time is required (on average) to answer with a lie.

There are limitations, however, to our methodology: the artificiality of the experimental setting, the low-risk nature of the lying involved and the fact that our subjects were mostly from academic backgrounds. There is still a need for more ecological studies of deception in humans.

Our finding of increased response time during lying is, however, congruent with a report of a convicted murderer, filmed while lying and telling the truth.[20] Although recounting similar material on both occasions, this subject exhibited slower speech with longer pauses and more speech disturbance when lying. He also exhibited fewer illustrators – less bodily movement.[20] Previous meta-analyses of behavioural lying studies have also pointed to speech disturbance, increased response latency and a decrease in other motor behaviours in the context of attempted deception.[20] Although responses on our computerised tasks were non-verbal, the behavioural and functional anatomical profile revealed may indicate a common process underlying these findings and others,[18,20] namely an inhibitory mechanism being utilised by those attempting to withhold the truth (a process associated with increased response latency).

It is noteworthy that the difference between lying and truth times for all groups in our studies was around 200 ms.[6,18] This is consistent with behavioural data from 'guilty knowledge' tasks performed with and without electrophysiological recording by other authors.[21,22]

Other groups have also used fMRI and found the prefrontal cortex to be implicated in deception. Although the foci reported may in some respects be different, the principle of engagement of executive brain regions seems to hold, as does the notion that 'truth' is a relative baseline.[23,24] None of these studies has reported areas of greater activation during truthful responding.[18,23,24]

Taken together, the data seem to support the contention that lying is an executive process, engaging higher brain regions notably within prefrontal cortex. They also seem to imply, however, that truth telling constitutes a relative baseline in human cognition, returning us to the conundrum of whether biology and ethics can always be regarded as totally independent domains.

References

1 Spence SA. Hysterical paralyses as disorders of action. *Cognit Neuropsychiatry* 1999;4:203–26.
2 Merskey H. *The analysis of hysteria: understanding conversion and dissociation*. London: Gaskell, 1995.
3 Ekman P, O'Sullivan M. Who can catch a liar? *Am Psychol* 1991;46:913–20.
4 Vrij A. *Detecting lies and deceit: the psychology of lying and the implications for professional practice*. Chichester: John Wiley, 2000.
5 *Chambers concise dictionary*. Edinburgh: Chambers Harrap, 1991.
6 Spence S, Farrow T, Leung D et al. Lying as an executive function. In: Halligan P, Bass C, Oakley P (eds), *Malingering and illness deception*. Oxford: Oxford University Press, 2003:255–66.
7 Dunbar R. On the origin of the human mind. In: Carruthers P, Chamberlain A (eds), *Evolution and the human mind: modularity, language and meta-cognition*. Cambridge: Cambridge University Press, 2000:238–53.
8 O'Connell S. *Mindreading: an investigation into how we learn to love and lie*. London: Doubleday, 1998.
9 Happe F. *Autism: an introduction to psychological theory*. Hove: Psychology Press, 1994.
10 Giannetti E. *Lies we live by: the art of self-deception*. Translated by J Gledson. London: Bloomsbury, 2000.
11 Ford CV, King BH, Hollender MH. Lies and liars: psychiatric aspects of prevarication. *Am J Psychiatry* 1988;145:554–62.
12 Spence SA, Hunter MD, Harpin G. Neuroscience and the will. *Curr Opin Psychiatry* 2002;15:519–26.
13 Badgaiyan RD. Executive control, willed actions, and nonconscious processing. *Hum Brain Mapp* 2000;9:38–41.
14 Jack AI, Shallice T. Introspective physicalism as an approach to the science of consciousness. *Cognition* 2001;79:161–96.
15 Shallice T. *From neuropsychology to mental structure*. Cambridge: Cambridge University Press, 1988.
16 Barnes TRE, Spence SA. Movement disorders associated with antipsychotic drugs: clinical and biological implications. In: Reveley MA, Deakin JFW (eds), *The psychopharmacology of schizophrenia*. London: Hodder Arnold, 2000:178–210.
17 Marnham P. The confidence man. *New York Review of Books*, 19 December 2002:66.
18 Spence SA, Farrow TF, Herford AE et al. Behavioural and functional anatomical correlates of deception in humans. *Neuroreport* 2001;12:2849–53.
19 Starkstein SE, Robinson RG. Mechanism of disinhibition after brain lesions. *J Nerv Ment Dis* 1997;185:108–14.
20 Vrij A, Mann S. Telling and detecting lies in a high-stake situation: the case of a convicted murderer. *Appl Cogn Psychol* 2001;15:187–203.
21 Farwell LA, Donchin E. The truth will out: interrogative polygraphy ("lie detection") with event-related brain potentials. *Psychophysiology* 1991;28:531–47.

22. Seymour TL, Seifert CM, Shafto MG, Mosmann AL. Using response time measures to assess "guilty knowledge". *J Appl Psychol* 2000;85:30–37.
23. Langleben DD, Schroeder L, Maldjian JA *et al*. Brain activity during simulated deception: an event-related functional magnetic resonance study. *Neuroimage* 2002;15:727–32.
24. Lee TMC, Liu H-L, Tan L-H *et al*. Lie detection by functional magnetic resonance imaging. *Hum Brain Mapp* 2002;15:157–64.

5

Values-based medicine: delusion and religious experience as a case study in the limits of medical-scientific reduction

BILL (KWM) FULFORD

Professor Bill (KWM) Fulford discusses the difficulty of differentiating delusion from religious experience according to current diagnostic protocols which seek to exclude value judgements. Analysis of this difficulty suggests the wider conclusion that, contrary to the medical-scientific expectation that technological advances in medicine will reduce the relevance of values-based assessment, they will in fact require its wider use. This is because scientific advances open up new choices in medicine and with choices go values. The future of medicine is thus not one of increasing reduction to science but rather of an enriched discipline in which science and values have equal and complementary roles.

I was once introduced by the president of a postgraduate medical dining club, an elderly retired surgeon-commander, who was obviously somewhat bemused by my hybrid job description. 'It's a pleasure to introduce Professor Fulford', he began. 'He tells me that he's a philosopher and a psychiatrist... He seems to have managed to combine *two* degenerate subjects in one!'

The origins of values-based medicine

The surgeon-commander's (entirely friendly) fire at my discipline, the philosophy of psychiatry, nicely captures the tensions implicit in the twin themes of this book, science and morality, as they apply to medicine.

Twentieth-century medicine, developing scientifically as never before, finally threw off traditional associations between sin and disease. Against this backdrop, psychiatry, with its still morally ambiguous diagnostic categories – psychopathy, alcoholism, the paraphilias (disorders of sexual object choice), and so forth – has the appearance, perhaps, of a pre-scientific discipline, an appearance which, on an exclusively scientific model of 20th-century medicine, is reinforced by psychiatry's continued engagement with philosophy.[1]

Yet there is a well-recognised paradox at the heart of the successes of 20th-century scientific medicine, namely that even as undreamt-of advances were being made in the prevention and cure of disease, so public unease with medicine actually increased. One response to this has been the emergence of a quasi-legal 'bioethics' to police the morals of medicine. Values-based medicine offers a radically different response: a response based on partnership rather than policing, and a response in which philosophy and psychiatry, the least of disciplines in 21st-century medical science, are pre-eminent.[2]

This chapter, then, is about integration. It is about the integration of science with morality in 21st-century medicine. More specifically, it is about the integration of evidence and 'values' as twin, increasingly necessary, but always mutually irreducible, partners in 21st-century medical decision-making.

The story of Simon

Let us start by considering a case history. Based on that of a real person,* I will call this the 'story of Simon'. I will outline a number of possible responses to Simon's story from the perspective of an exclusively scientific model of medicine. The failure of these responses, as judged from Simon's perspective, will indicate the need for a new model, the model of values-based medicine, in which science and values are fully combined.

> *Simon was an African-American working in a racist area of the USA as a successful lawyer. He came from a Baptist family but was not a particularly religious man. He had had occasional contact with seers (I gather that this is not at all unusual in America). So we have this ordinary, straightforward, middle class black lawyer working in a rather dangerous area. Then, at the age of 40, Simon was suddenly threatened with legal action by a group of his white colleagues. As a racially motivated attack, this was a threat to his professional and personal identity, to his continued prosperity, and, possibly, even to life and limb. So it was a very serious threat.*
>
> *Simon reacted to all this in a rather dramatic way. He went home one evening and set up an altar in his front room. This took the form of a table on which he placed two candles, and between the candles, a big family bible. Then Simon, who you will remember was not a particularly religious man, knelt in front of this altar and spent the whole night praying.*

---

*Simon's story is one of a series collected by the psychologist Mike Jackson as part of his DPhil studies in Oxford.[2,3] The links between these studies and philosophical value theory were first published by Jackson and Fulford.[4]

> When he got up in the morning he found that wax from the candles he had put on his altar had run down on to the family bible outlining certain words and phrases. Seeing these 'seals', as he called them, Simon knew immediately that 'something remarkable was going on'. He called a friend, saying 'I saw the seal, which is in my father's bible, and you know, something remarkable is going on over here. It is the beauty of it that is shining through... It is not my mind that is playing games because it is the specificity of the words that is telling me what to do.'

Delusion or religious experience?

Those were Simon's actual words. If a psychiatrist or mental health worker saw a patient with this history, they would diagnose it as a delusion. Yes, it could be a delusion. And it could be something even more specific, a *delusional perception* – that is, a delusional set of ideas springing from an otherwise normal perception. And this was how it was for Simon. The wax seals told him that he was the new Messiah and that he was going to reconcile Islam with Christianity. They were very specific to Simon. The rest of his social and professional group could see no particular significance, still less specific messages, in his 'seals'. But for Simon they were the most important thing that had ever happened to him.

So how did the story end? Did Simon win his court case? Or finish up in hospital? In a word, was this an empowering (if idiosyncratic) religious experience, or a delusion, a symptom of a severe mental illness?

Well, by the criteria of traditional psychiatry, there is no doubt that Simon should have ended up in hospital. He had not just a delusion but a delusional perception. (This is considered in more detail below.) And in traditional psychiatry, a delusional perception is pathognomonic of schizophrenia, or of some other closely related severe psychotic disorder.*

But – and here is the rub for traditional psychiatry – Simon was not ill at all! Idiosyncratic as his experiences were, they actually guided and empowered him. He took on his (highly dangerous) court case. He won it. His stock as a lawyer, in this racist area of America, went sky high. The last we heard of him (from Oxford), he was setting up a foundation for the study of religious experience.

---

*Recent research shows that delusional perception and other symptoms originally thought to be pathognomonic of schizophrenia may occur in other psychotic conditions. Psychiatry differentiates these conditions by related features such as dementia by impaired memory and so on, mania by elevated mood, flight of ideas etc. Simon's story makes an excellent test case for differential diagnosis with trainee psychiatrists.

From symptoms to diagnosis in psychiatry

So far this may look like, in tennis scoring, fifteen–love against psychiatry. So how might psychiatry even up the score? One response, from the perspective of traditional psychiatry, is to contest the score, to argue that Simon did not, really, satisfy the criteria for a diagnosis of schizophrenia (or some related psychotic disorder) if these criteria are sufficiently rigorously applied.[5]

Simon's experience, however, really does conform to the psychopathological 'delusional perception'. This is important because psychiatry, modelling itself (appropriately) on physical medicine, has attempted to resolve issues of differential diagnosis, in the first instance, by the scrupulously careful definition and description of symptoms. In heart disease, for example, genuine heart pain is distinguished from other complaints of pain in the chest by specific phenomenological features – central chest pain, crushing, radiating into the throat, down the left arm and so on. The first lesson I learnt as a medical student was that a patient who points to his heart when asked to indicate the location of the pain is unlikely to have heart pain.

Similarly in psychiatry, then, the way to distinguish genuine symptoms from the range of other unusual experiences to which people are subject is by careful observation.

This well tried and tested approach was the basis of what is widely regarded as one of the gold standards of medical scientific psychiatry, the Present State Examination or PSE.[6] This was developed as a research instrument over 10 years at the Institute of Psychiatry in London. In the PSE a delusional perception is said to be 'based on a sensory experience and involves suddenly becoming convinced that a particular set of events has a special meaning'. This would certainly fit Simon's case, and his experience, rated by a researcher skilled in the use of the PSE, confirms its status as a delusional perception so defined.[3] So Simon's experience was not just an odd or unusual experience. It had the specific characteristics of a delusional perception.

The next question is what does this symptom mean for psychiatric diagnosis? For this, consider another gold standard of psychiatric diagnosis, the World Health Organization's (WHO) diagnostic manual, the *International Classification of Disease (ICD)*.[7] The *ICD* defines schizophrenia (rather as conditions such as migraine are defined) symptomatically. According to the *ICD*, then, for a diagnosis of schizophrenia you have to have, among other things, at least one symptom from a given list. This list includes 'delusional perception'. Although I have given you only part of Simon's story he also had another relevant symptom from the *ICD* list, called 'thought insertion'. Now what all this means is that by the lights of the best of scientific psychiatry, as represented by the PSE and the *ICD*, Simon has not one but two 'votes' for a diagnosis of schizophrenia or some related psychotic disorder. So he was not just ill: he had (or ought to have had) a severe mental illness – and yet as we have seen, not only was Simon not ill but he was very much empowered by his experiences.

So even on closer scrutiny, after (to extend the tennis metaphor) checking with the referee (scientific psychiatry), it looks as though the score of fifteen–love against psychiatry has to stand. But hold on: psychiatry is calling in the umpire!

A second look at psychiatric diagnosis

The umpire in this case is the main competitor to the WHO's *ICD*, a classification produced by the American Psychiatric Association (APA) called the *Diagnostic and Statistical Manual*, or *DSM*.[8] The current edition is *DSM-IV*. The *DSM-IV* task force of the APA worked hard to coordinate its work with the *ICD* group in the WHO. But whereas the priority for the WHO was to produce a classification which was acceptable worldwide, and hence across a number of different cultures, the DSM could focus more exclusively on the scientific evidence supporting diagnosis in psychiatry.

It is worth looking at the introduction to *DSM-IV*. The very first page is full of the language of evidence-based medicine. The *DSM-IV*, its authors emphasise, will be based on 'an extensive empirical foundation'; hence they 'took a number of precautions to ensure that [diagnostic categories] would reflect the breadth of available evidence'; they thus 'established a formal evidence-based process', including international representation to secure 'the widest pool of information'.[8]

What, then, does the *DSM*, this most scientific of psychiatric diagnostic classifications, suggest in Simon's case? Surprisingly perhaps, given the extent of the collaboration between the *ICD* and *DSM* task forces, *DSM* suggests a really *crucial* difference in the way that the psychotic disorders are diagnosed. The *DSM* has two main sets of criteria for a diagnosis of schizophrenia, A and B. The first of these, criterion A, is a list of symptoms. Essentially, this is the same list as that in the *ICD*. So, to the extent of criterion A, *DSM* and *ICD* give the same diagnosis for Simon. But then comes the crucial difference: criterion B, which is also essential for a diagnosis of schizophrenia in *DSM*, does not appear at all in *ICD*.

Criterion B is one of 'social/occupational dysfunction'. It reads 'For a significant portion of the time since the onset of the disturbance one or more major areas of function such as work, interpersonal relations or self-care, are markedly below the level achieved prior to the onset'. So, what does this suggest about

Simon? Was his occupational functioning reduced? Well, obviously not. It was actually increased. So a diagnosis of schizophrenia, which seemed inevitable with the *ICD*, is flatly contradicted by the more 'scientific' *DSM*.

The *DSM*, then, escapes from the diagnostic choice of 'schizophrenia' to which the WHO's *ICD* led us. And this appears, at first glance, to be good news for scientific psychiatry. For the *DSM*, remember, is more explicitly evidence-based than the *ICD*. So the more uncompromisingly science-based *DSM* gives us what the *ICD* could not, a diagnosis with, in Simon's case, considerably more face validity. The *DSM* suggests, consistently with the *prima facie* facts in Simon's case, that he is not ill. Yes, he does have a psychotic experience. Yes, this experience has the form of a delusional perception. But, no, he does not show (on the evidence to date) the additional feature required by *DSM* of impaired social/occupational functioning. Hence by the criteria in *DSM*, Simon does not have schizophrenia.

Psychiatric science, then, called in at the higher level of the umpire (the *DSM*), reverses the decision of the lower-level referee (the *ICD*) and evens up the score at fifteen all.

Value judgements in psychiatric diagnosis

But now notice this. There is something distinctly odd about criterion B from the point of view of medical science. The point is this. Criterion B is about change in functioning. So far so good. *Change* in functioning is indeed something that scientific means might be used to measure; after all we use scientific means to measure changes in bodily functioning (of the heart, for example) in physical medicine. But criterion B involves more than a change in functioning; it involves a change in functioning for the *worse*. And this takes us beyond the neutral measurements of science into the realm of *value judgements*.

The value judgements involved in criterion B become obvious if we emphasise the key words and phrases in which the criterion is framed. To satisfy criterion B, the patient's social/occupational functioning must be not merely different but **worse**; that is to say, it must be '*below* the level' previously achieved. In an extension of the criterion to cover children, the *DSM* goes on to specify that there must be a '*failure* to achieve the expected level'.

So with criterion B in *DSM*, a whole series of value judgements are found where, from the perspective of the traditional view of medical science, no value judgements should be.

The significance of these value judgements will be investigated in more detail later in this chapter. But, to anticipate a little, that criterion B *is* crucial diagnostically involves something of a paradigm shift. For it brings value judgements right into the heart of what has previously been perceived as an area of medicine reserved exclusively to medical science, *viz* diagnosis. In being concerned with psychotic disorders, furthermore, criterion B brings in value judgements not at the edge of medical psychiatry, in relation to the overtly ambiguous concepts of the kind noted at the start of this paper, such as psychopathy, alcoholism and the paraphilias, but at the very heart of traditional descriptive psychopathology. Furthermore, in being a part of the *DSM*, criterion B brings in value judgements not in some dubious diagnostic system on the fringes of scientific medicine but at the heart of the most self-consciously evidence-based of psychiatry's classifications.

Time out

The rest of this chapter will ask how to interpret the fact that value judgements are appearing at the heart of psychiatric diagnosis. It will consider three perspectives; that of the moral pole, that of the science pole and a 'third way'.

Interpretations from the first (moral) and second (scientific) perspectives draw on the traditional model of medicine as being essentially a science. It is within this model that psychiatry, to echo the elderly surgeon-commander introduced at the start of this paper, has been stigmatised as a 'degenerate' subject. It is degenerate in different ways from the two perspectives, but it is degenerate nonetheless.

The third interpretation, the third way, draws on that other (in the elderly surgeon-commander's view) degenerate discipline, philosophy. However, philosophy, in this case, far from undermining scientific medicine, clarifies its proper role. This in turn, as we will see, will clarify the proper role of value judgements in medicine. And this will lead us, finally, to the concept of values-based medicine (VBM). VBM, as will be briefly outlined, is a set of knowledge and clinical skills that stand to the value judgements involved in medicine, much as evidence-based medicine stands to the facts. This is why, as already indicated, VBM offers a partnership model in place of the policing model of quasi-legal ethics. The end point then will be a view of medicine in which fact and value, science and morality, have equal and complementary roles.

Interpretation of the value judgements in Simon's case from the moral pole

A brilliantly clear exemplar of the 'moral' interpretation of the appearance of value judgements at the heart of psychiatric classification and diagnosis is to be found in the work of the American psychiatrist – and *bête noire* of scientific psychiatry – Thomas Szasz.

Szasz came to prominence in the 1960s, shortly after being made full Professor of Psychiatry at Syracuse University, with a book provocatively titled *The Myth of Mental Illness*.[9] Szasz argued that the value-laden nature of psychiatric diagnostic concepts compared with their physical medicine counterparts pointed to a difference in kind between mental disorders and bodily diseases. Diseases, he suggested, are defined by scientific norms, the norms of anatomy and physiology. Mental disorders on the other hand are defined by value norms, norms which are 'psychosocial, ethical and legal' in character. Mental disorders, then, are not part of the scientific world of medical diseases at all. Properly understood, they are moral or, as Szasz called them, 'life' problems.

Interpretation of the value judgements in Simon's case from the science pole

A second interpretation of the appearance of value judgements, as in Simon's case, at the heart of psychiatric diagnosis, is that psychiatry is at a preliminary, or perhaps primitive, stage of its development as a medical science.

RE Kendell, who went on to become a distinguished President of the Royal College of Psychiatrists, authoritatively argued that this was not the case in 1975.[10] Drawing on a wide-ranging and scholarly analysis of the many different interpretations of the concept of disease, Kendell argued, in effect, that where values appear in psychiatric classification and diagnosis, they can be re-defined (in principle) in terms of the facts of evolutionary biology.

The impairment of function required by criterion B, then, although apparently a matter of value judgements, can, according to Kendell's argument, be redefined in terms of 'biological dysfunction', this in turn being defined by reference to the facts specifically of reduced evolutionary fitness, *viz* reduced potential for survival and/or reproduction. And biological dysfunction, so defined, Kendell (and others)* have assumed, is a matter, in principle, of evidence. Kendell indeed drew on epidemiological evidence to show that a number of mental disorders are associated with reduced life and/or reproductive expectations.

---

*Kendell took his definition of disease from work in respiratory medicine;[15] the 'biological dysfunction' approach was developed further by the American philosopher Christopher Boorse,[16] drawing on the distinction between disease and illness and applying it to mental illness.[17] Boorse has subsequently presented a more detailed account of his model.[18] The American social worker and philosopher Jerome Wakefield has produced a spirited defence of this model with *DSM* particularly in mind. For recent philosophical moves in this debate, including articles by Szasz, Wakefield and Fulford, also Chris Megone and Tim Thornton, see a special issue of the journal, *Philosophy, Psychiatry, & Psychology*.[19]

The first and second interpretations considered together

There is much to be said for and against both Kendell and Szasz's interpretations. Both are carefully argued and scholarly. Both, it is true, are vulnerable to many counter-examples.[11] But Szasz's position, that mental disorders are, really, moral not medical problems, is supported by the vulnerability (to say the least) of psychiatry to being used as a means for controlling socially unwelcome behaviour. In the USSR, political dissidents were diagnosed as suffering from 'delusions of reformism' and treated in mental institutions.[12] A religious group, the Falun Gong, face similar 'diagnoses' in China at the present time.[13] And in the UK, many fear that proposed new mental health legislation will equate dangerous behaviour with mental disorder.[14] In defence of a medical model of mental disorder, on the other hand, Kendell could point to progress in drug treatment, and, with recent advances in the neurosciences (in particular brain imaging and behavioural genetics), to the emergence of credible disease models of conditions such as schizophrenia.

If we step back for a moment from the content of these arguments, however, and instead think about their structure, we notice something rather surprising. For Szasz and Kendell, as shown in the first chapter of my *Moral theory and medical practice*,[20] although ostensibly disagreeing about the status of mental illness, are really disagreeing about the meaning of *physical* illness. Szasz defines physical illness in terms of anatomical and physiological norms. Kendell defines it in terms of evolutionary norms of reduced life and/or reproductive expectations. Hence Kendell's epidemiological evidence is nothing to Szasz, and Szasz's absence of anatomical and physiological norms is nothing to Kendell.

What is required, then, for the game to move forward, is a deeper analysis of the concept of physical illness. It is this that is offered by philosophical value theory.

Third interpretation – philosophical value theory

Philosophical value theory, drawing particularly on the work of a former White's Professor of Moral Philosophy in Oxford, RM Hare,[21] offers an interpretation of the more value-laden nature of mental illness which allows us to have our cake, philosophically speaking, and to eat it. To see this we need to dip briefly into Hare's analysis of the relationship between evaluative and descriptive (or factual) meaning.

The key point to take from Hare's work in this context is that where value terms are used to express values which are largely *shared*, they can come to look like *factual* (or descriptive) terms. Hare gives the example of good strawberry: etymologically this is a value term; yet in everyday discourse it conveys the factual meaning along the lines of a strawberry that is a red, sweet, juicy, grub-free strawberry.

The reason for this is obvious enough, Hare continued. For value judgements are made on the basis of criteria which are descriptive or factual in nature – 'red, sweet, juicy and grub-free' all describe, or state facts about, a strawberry. Hence, to say that our values are shared is just to say that we share the same *factual* criteria for the value judgement in question. Hence these criteria become stuck by association to the use of that value term. In the case of strawberries, most people judging a strawberry to be good have in mind that it is, as a matter of *fact*, 'red, sweet, juicy and grub-free'. By simple association, therefore, the *factual* meaning 'red, sweet, juicy and grub-free' becomes stuck to the meaning for the value term 'good' when used of strawberries.

Hare, writing in the 1950s, was not thinking specifically of the language of medicine. He was concerned rather with the logical properties of value terms in general. But his 'good strawberry' example gives us all we need for a deeper analysis of the meaning of physical illness and hence for a third way between the 'values-in' versus 'values-out' extremes of the polarised science versus morality positions in the debate about mental illness. This is because, as Figure 5.1 illustrates, Hare's analysis of the way in which value terms in general can come to look like factual terms maps directly on to the medical case.

Thus, medical terms – 'illness', 'disease', dysfunction', and so forth – like the term 'good strawberry' (in the left-hand side of Figure 5.1), are etymologically value terms (*ill*ness, *dis*ease, *dys*function are negatively evaluated counterparts of such terms as *well*ness, *health* and *good* function). Yet, also like 'good strawberry', these medical terms carry largely factual connotations, at least as they are used in the context of physical medicine. And this makes sense in the Hare model, because physical medicine is concerned with aspects of human experience and behaviour over which, as with good strawberries, our values are largely shared – a 'heart attack', for example, involving as it does severe physical pain and imminent risk of death, is, in and of itself, a bad condition by (nearly) anyone's standards.

According to the Hare model, then, terms such as disease, illness and dysfunction appear value-free when they are used in physical medicine, not because they are purely factual but because the values they express in such contexts are largely shared. There is no criterion B for a 'heart attack', then, not because diagnosis in cardiology is value-free but because the value judge-

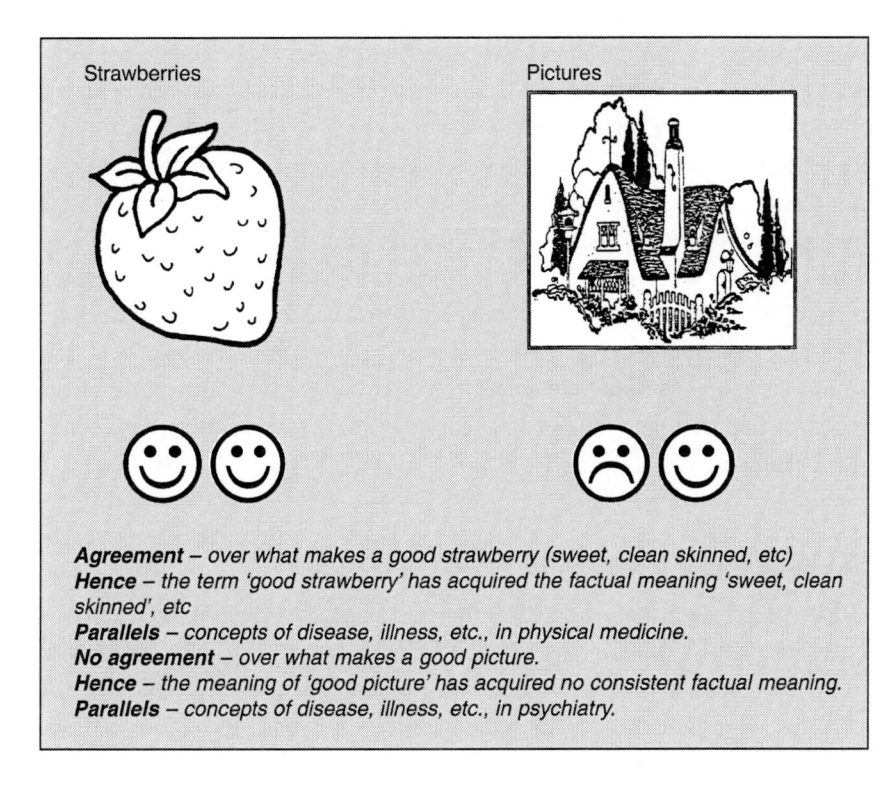

Figure 5.1 Factual and evaluative meaning. Philosophical value theory shows how a value term such as 'good strawberry' can come to look like a factual term where, as in the case of strawberries, the factual criteria for the value judgement expressed by the value term in question are widely shared. Terms such as 'disease' and 'illness' then, according to this theory, are mainly factual in meaning in physical medicine, and more value-laden in psychiatry not because physical medicine is more scientific than psychiatry, but because psychiatry is concerned with areas of human experience and behaviour (such as emotion, desire, volition and belief) in which human values are particularly diverse.

Strawberries

Pictures

Agreement – over what makes a good strawberry (sweet, clean skinned, etc)
Hence – the term 'good strawberry' has acquired the factual meaning 'sweet, clean skinned', etc
Parallels – concepts of disease, illness, etc., in physical medicine.
No agreement – over what makes a good picture.
Hence – the meaning of 'good picture' has acquired no consistent factual meaning.
Parallels – concepts of disease, illness, etc., in psychiatry.

ments involved in defining poor heart functioning are widely shared and hence unproblematic. Another way of putting this is to say that there is indeed in *principle* a criterion B for heart attack; but since there will never be any disagreement over its application in *practice*, it can safely be ignored and hence it is left implicit.

By contrast, therefore, terms such as mental illness and mental disorder are relatively value-*laden*, not because of a lack of scientific content (as in the Kendell model) and still less because they are moral as distinct from medical concepts (as in the Szasz model), but because the values they express are relatively open and contentious. Hare gave examples such as 'good poem' and 'good picture' for value judgements over which people's values are highly divergent and hence which are value-laden counterparts to 'good strawberry'. The parallels here are illustrated in the right-hand side of Figure 5.1. As with 'good picture', then, psychiatry is concerned with areas of human experience and behaviour over which our values are highly divergent – emotion, desire, volition, belief, sexuality and so forth are all areas in which people's values differ widely and legitimately.

Notice the economy of this account. Drawing on no more than a general logical property of value terms, together with a contingent feature of human values, Hare's model explains even-handedly the more value-laden nature of mental illness and the more fact-laden nature of physical illness.* It explains, *contra* Kendell, why mental illness is genuinely more value-laden than physical illness (rather than this being provisional on future scientific advances). But, *contra* Szasz, the explanation from philosophical value theory shows mental illness to be more value-laden than physical illness *just in being* a genuine species of illness (rather than a moral or life problem). To this extent, then, philosophical value theory provides us with a theoretical framework for reconciling the moral and scientific sides of medicine within a balanced view of the discipline as a human science. It is to an outline particularly of the practical implications of this balanced view that I turn in the remainder of this chapter.

Values and evidence in clinical decision-making

Values-based medicine, as noted at the start of this chapter, is like evidence-based medicine (EBM) to the extent that both are responses to complexity. EBM has come in for much criticism: it has been accused (perhaps not unjustly) of prioritising highly generalised information derived from high-quality research at the expense of individual experience. Such criticisms should be understood as criticisms not of EBM as such but as criticisms of an inadequate, or perhaps incomplete, EBM. So understood, then, just as EBM is a response to the growing complexity of the facts bearing on decision-making in healthcare, VBM is a corresponding response to the growing complexity of the values bearing on decision-making in healthcare.

Bioethics, as also noted in the introduction to this paper, is one important response to the growing complexity of values in healthcare. But bioethics, in the strongly quasi-legal form it takes in practice, assumes shared values, 'right' outcomes against which medical and other decisions in healthcare can be assessed ethically. These right values are incorporated into codes, guidelines and other forms of regulation. But as Simon's case illustrates, at least in mental health, many of the problems with which we are concerned, even in diagnosis, are defined by values which, far from being shared, are widely and legitimately *diverse*.

---

*I develop the details of the theory supporting this 'third way' in *Moral theory and Medical Practice*:[20] this includes an analysis of the relationship between the patient's experience of illness and medical knowledge of the causes of disease (Chapter 4), and a detailed account of the applications of philosophical value theory to particular kinds of psychopathology (Chapters 8–10). I extend the theory in an article from 2000.[22]

Philosophical value theory, then, suggests that although quasi-legal bioethics is important in some contexts, we need a wider range of 'tools' if we are to respond adequately to the growing complexity of values in healthcare. Codes and guidelines, according to philosophical value theory, are appropriate for values that are largely shared within a given group. As such, they provide an important framework for practice. By the same token, though, codes and guidelines will be inappropriate where legitimately different values are in play. For in such situations, the values expressed by the code or guideline will be in conflict with the values of many of those to whom the rules are intended to apply.

VBM: from legal/ethical rules to improved clinical standards

VBM, then, shifts the emphasis from rules embodying particular values that prescribe 'right outcomes', to clinical skills supporting 'good process' in healthcare decision-making. A training manual in the skills base of VBM, *Whose values?*, was launched by Rosie Winterton, as the Minister of State in the Department of Health with responsibility for mental health, at a conference in London in 2004.[23] The skills-base of VBM, as *Whose values?* describes, is fourfold – awareness, knowledge, reasoning skills and communication skills, as described below.

Raising awareness

New methods for raising awareness of values were developed originally by Warwick University Medical School's programme in the Philosophy and Ethics of Mental Health (PEMH) and the Sainsbury Centre for Mental Health (SCMH), in London. The methods were piloted and shown to be effective in such challenging areas as assertive outreach and community mental health.[24]

Increased knowledge

A range of both empirical and philosophical methods are available for increasing knowledge and values.[25] A study completed at Warwick University showed the power of research protocols combining philosophical and social science methods to reveal the often very divergent values of different stakeholders (users, carers, nurses, social workers and psychiatrists) in healthcare.[26]

Reasoning skills

As in quasi-legal ethics, principles, casuistry and any other method of ethical reasoning may be helpful in VBM. The difference, though, is that in quasi-legal ethics, ethical reasoning is used primarily to tell us which values are 'right', whereas in VBM ethical reasoning is used primarily to explore differences of values.[27,28]

Communication skills

Communication skills also have different roles in VBM and in quasi-legal ethics. In quasi-legal ethics their role is 'executive'– that is, to help in *applying* the rules. In VBM communication skills have a *substantive* role: they are central both to understanding the values of those concerned in a given situation and to resolving conflicts and difficulties arising from differences of values. This connects VBM with a model of medical student education in which ethics, law and communication skills are combined in a clinical problem-solving approach to decision-making.[29]

VBM: from theory into policy and practice in psychiatry

Building on the training manual, *Whose Values?*, there have been a number of national and international policy and service development initiatives in values-based practice for psychiatry and related areas of mental health and primary care.

A key building block for these initiatives in the UK is a National Framework of Values that has been adopted by the National Institute for Mental Health in England (NIMHE).[30] The Framework, which is reproduced in Box 5.1, starts with three central principles of values-based practice.

The principle of 'respect', emphasising as it does the central importance of starting from the values of each individual patient, may sound excessively relativistic. However, the Framework goes on to set out a number of specific policy constraints that are imposed by the values-based approach. Important among these is the negative constraint that any form of discrimination is ruled out by values-based practice. This is because discrimination in all its forms is incompatible with the principle of respect for diversity that is at the heart of the values-based approach. But there are also a number of positive constraints, including the importance of multidisciplinary teamwork, of recovery, of strengths-based assessment, of a dynamic and flexible approach.

NIMHE is the section of the Care Service Improvement Partnership in the Department of Health in London directly responsible for delivering national policy in mental health and social care. Correspondingly, the values-based approach, building on the NIMHE Framework of Values, has been incorporated into a number of national policy and service development initiatives.

> **Box 5.1. NIMHE National Framework of Values**
>
> *The National Framework of Values for Mental Health*
>
> The work of the National Institute for Mental Health in England (NIMHE) on values in mental health care is guided by three principles of values-based practice:
>
> **Recognition** – NIMHE recognises the role of values alongside evidence in all areas of mental health policy and practice
>
> **Raising Awareness** – NIMHE is committed to raising awareness of the values involved in different contexts, the role/s they play and their impact on practice in mental health
>
> **Respect** – NIMHE respects diversity of values and will support ways of working with such diversity that makes the principle of service-user centrality a unifying focus for practice. This means that the values of each individual service user/client and their communities must be the starting point and key determinant for all actions by professionals.
>
> Respect for diversity of values encompasses a number of specific policies and principles concerned with equality of citizenship. In particular, it is anti-discriminatory because discrimination in all its forms is intolerant of diversity. Thus respect for diversity of values has the consequence that it is unacceptable (and unlawful in some instances) to discriminate on grounds such as gender, sexual orientation, class, age, abilities, religion, race, culture or language.
>
> Respect for diversity within mental health is also:
>
> - *User-centred* – it puts respect for the values of individual users at the centre of policy and practice
> - *Recovery oriented* – it recognises that building on the personal strengths and resiliencies of individual users, and on their cultural and racial characteristics, there are many diverse routes to recovery
> - *Multidisciplinary* – it requires that respect is reciprocal, at a personal level (between service users, their family members, friends, communities and providers), between different provider disciplines (such as nursing, psychology, psychiatry, medicine, social work) and between different organisations (including health, social care, local authority housing, voluntary organisations, community groups, faith communities and other social support services)
> - *Dynamic* – it is open and responsive to change
> - *Reflective* – it combines self monitoring and self management with positive self regard
> - *Balanced* – it emphasises positive as well as negative values
> - *Relational* – it puts positive working relationships supported by good communication skills at the heart of practice.
>
> NIMHE will encourage educational and research initiatives aimed at developing the capabilities (the awareness, attitudes, knowledge and skills) needed to deliver mental health services that will give effect to the principles of values-based practice.

Key to these is the '10 Essential Shared Capabilities'[31] (10 ESCs), a training initiative in the generic skills required for anyone working in mental health and social care, and indeed other areas of primary care, that is built on both evidence-based and values-based sources. A core set of training materials has been produced to support the 10 ESCs and further more specific modules, for example, Delivering Race Equality[32] and for Community Development Workers,[33] are currently being developed. The 10 ESCs in turn supports policies on new roles and ways of working for psychiatrists and for related mental health professions[33] which underpin the National Workforce Strategy for Mental Health,[34] this in turn being the foundation for the National Service Framework for Mental Health[35] and recent updates such as the Priorities and Planning Framework.[36]

In addition to these NIMHE-led initiatives, values-based approaches are being developed to support new legislation, including the Mental Capacity Act and the proposed amendment to the Mental Health Act 1983. The approach is also relevant to areas such as audit and commissioning, and in this context is explicitly part of the Welsh Assembly policy, Healthcare Standards for Wales.[37] Although led from the UK, there have been training initiatives in values-based practice in a number of other countries, including Brazil, Belgium, Holland, Portugal, South Africa and Turkey. There is also an active research field, both within the UK (supported particularly by NIMHE, by Warwick Medical School and by the Mental Health Foundation in London), and within the wider international development of philosophy of psychiatry, through the International Network for Philosophy and Psychiatry, the Association of European Psychiatrists and the World Psychiatric Association.[14]

VBM: from psychiatry to bodily medicine

In this chapter I have concentrated on the development of VBM in psychiatry. We need VBM in psychiatry, philosophical value theory suggests, because of the complexity of the values bearing on clinical decision-making in the areas of human experience and behaviour with which psychiatry is particularly concerned – that is, emotion, desire, volition, sexuality and so forth. The fact that psychiatry is relatively value-laden compared with physical medicine does not make it, to use the word of my surgeon-commander at the start of the chapter, 'degenerate' scientifically. On the contrary, it is more complex evaluatively. That is, in that psychiatry differs from most areas of bodily medicine in being concerned with areas of human experience and behaviour in which human values are particularly diverse.

Philosophical value theory also suggests that with future scientific progress, VBM will become increasingly important not only in psychiatry but also in the rest of medicine. This is because scientific progress opens up choices, and with choices come values.

That scientific progress should increase the importance of values in medicine may be counter-intuitive from the perspective of a traditional, exclusively scientific, understanding of science. But we are already witnessing science causing an increase in the complexity of values in reproductive medicine, for example. Twenty years ago reproductive medicine was concerned with major pathology, such as the 'impacted fetus' and infertility, conditions which, like heart attacks, are 'bad' conditions on almost anyone's scale of values. But now a remarkable series of advances in such areas as genetics and endocrinology have given us choices in areas that until recently were the stuff of science fiction, such as assisted reproduction, fetal selection, reversing the menopause – even genetic selection is a realistic possibility. With these new choices, then, the key clinical variables in reproductive medicine are fast becoming emotion, desire, volition, belief and so forth. These are just the areas in which (as in psychiatry) human values differ widely and legitimately.

In reproductive medicine, then, so increasingly in other areas of bodily medicine, scientific and technological progress, by opening up new choices, will increasingly make it essential that VBM stands alongside EBM as a full partner in clinical decision-making.*

Conclusions: from scientific reduction to an enriched science and values model of medicine

It seems we have come a long way from the 'two degenerate subjects', the twin targets of the surgeon-commander's joke at the start of this chapter.

---

*EBM and VBM are thus a good deal closer than the traditional stand-off between science and values might suggest. There are also differences, of course. A key difference is that where EBM is based on objective knowledge, VBM is based on subjective understanding. The aim of science, as the American philosopher Thomas Nagel[38] put it, is a view from nowhere. This is its strength. It is also the justification for placing meta-analyses of RCTs at the top of the 'evidence hierarchy' in EBM (Sacket[39]). Such meta-analyses, in seeking a view from everywhere hope to approximate to the view from nowhere. But VBM, in being concerned with human values, aims to get as close as possible to the particular value perspectives of those concerned in a given situation. Again, there is no conflict here. The *a*perspectival ambitions of EBM and the *perspectival* ambitions of VBM provide complementary resources for clinical decision-making. In their differences, then, as in their similarities, VBM and EBM are partners in clinical decision-making.

Philosophy, one of his targets, has given us a model of medicine, the model of values-based medicine or VBM, a model in which, contrary to the traditional opposition of values and science, values and facts are fully integrated in clinical decision-making. VBM, moreover, has also emerged from the surgeon-commander's other target, psychiatry. And it is a specific prediction of VBM, guided by philosophical value theory, that scientific progress, through opening up an ever wider range of choices in healthcare, will increasingly cause the full diversity of human values to be brought into all areas of medicine. Far from being a degenerate end of scientific medicine, then, psychiatry, in being first into VBM, is showing us the future of scientific medicine as a whole.

There is of course a good deal more to the science/morality debate than the simple opposition between fact and value.* There is a good deal more to this debate even in medicine. A whole series of philosophical dualisms – freewill and determinism, agency and functioning, explanation and understanding, causes and meanings, the deep problem of mind and brain itself – are all just below the surface of everyday medical practice. This is perhaps one reason why medicine is so often so peculiarly difficult. The problems of philosophy, which as theoretical problems have baffled the brightest and the best for over two thousand years, have to be faced in medicine as matters of everyday practice.

Small wonder, then, that medicine has sometimes opted for the intellectual equivalent of the quiet life, a positivist model in which our essentially *human* discipline is reduced to a narrow biologism, sub-human, mechanistic and scientistic. Small wonder too that psychiatry, the area of medicine most resistant to such reductive simplifications, the area in which medicine's implicit philosophical complications come closest to the surface, should have been stigmatised as degenerate. But the fact/value element of the science/morality debate illustrates the dangers of reductionism in medicine. It also shows the strengths of anti-reductionism. For anti-reduction in medicine is the basis for an enriched fact and values model of clinical decision-making that is both fully science-based and fully patient-centred.

---

*Many philosophers have attacked the very distinction between fact and value. In this they continue an important and still unresolved debate stretching back at least to the 18th century empiricist philosopher, David Hume.[40] A particular criticism of philosophical value theory, as exemplified by the work of RM Hare, is that it is impotent practically (eg Williams[41]). Medicine shows this criticism to be misplaced. Like mathematics, philosophical value theory, just in being an analytic discipline, has to be applied in a practical context if it is to have practical effect. But also like mathematics, it is its status as an analytic discipline that makes philosophical value theory a particularly effective tool when it is so applied.

References

1. Fulford KWM. Philosophy meets psychiatry in the twentieth century – four looks back and a brief look forward. In: Louhiala P, Stenman S (eds). *Philosophy meets medicine.* Helsinki: Helsinki University Press, 2000.
2. Jackson MC. *A Study of the Relationship between Spiritual and Psychotic Experience.* Unpublished D.Phil thesis: Oxford University, 1991.
3. Jackson MC. Benign schizotypy? The Case of Spiritual Experience. In GS Claridge (ed). *Schizotypy: relations to illness and health.* Oxford: Oxford University Press, 1997.
4. Jackson M. and Fulford KWM. Spiritual experience and psychopathology. *Philos Psychiatr Psychol* 1997;4:41–66. Commentaries by Littlewood R, Lu FG *et al*, Sims A and Storr A and response by authors, pp67–90. Reprinted as Jackson, MC and Fulford, KWM (2002). Spiritual experience and psychopathology. In Fulford KWM, Dickenson D and Murray TH (eds) *Healthcare ethics and human values: An introductory text with readings and case studies.* Oxford: Blackwell Publishers: pp141–49.
5. Sims A. Commentary on 'Spiritual experience and psychopathology'. *Philos Psychiatry Psychol* 1997;4:79–82.
6. Wing JK, Cooper JE, Sartorius N. *Measurement and classification of psychiatric symptoms.* Cambridge: Cambridge University Press, 1974.
7. World Health Organization. *The ICD-10 classification of mental and behavioural disorders: clinical descriptions and diagnostic guidelines.* Geneva: World Health Organization, 1992.
8. American Psychiatric Association. *Diagnostic and statistical manual of mental disorders (fourth edition).* Washington DC: American Psychiatric Association, 1994.
9. Szasz TS. The myth of mental illness. *Am Psychol* 1960:15;113–18.
10. Kendell RE. The concept of disease and its implications for psychiatry. *Br J Psychiatry* 1975:127:305–15.
11. Fulford KWM. Mental illness, concept of. In: Chadwick R (ed). *Encylopedia of applied ethics.* San Diego: Academic Press, 1998.
12. Bloch S, Reddaway P. *Russia's political hospitals: the abuse of psychiatry in the Soviet Union.* London: Gollancz, 1997. Also published in the USA as *Psychiatric terror.* New York: Basic Books, 1997.
13. Human Rights Watch/Geneva Initiative on Psychiatry. *Dangerous minds: political psychiatry in china today and its origins in the mao era.* New York: Human Rights Watch, 2002.
14. Fulford KWM, Morris KJ, Sadler JZ, Stanghellini G. Past improbable, future possible: the renaissance in philosophy and psychiatry In: Fulford KWM, Morris KJ, Sadler JZ, and Stanghellini G (eds). *Nature and narrative: an introduction to the new philosophy of psychiatry.* Oxford: Oxford University Press, 2003.
15. Campbell EJ, Scadding JG, Roberts RS. The concept of disease. *BMJ* 1979;2: 757–62.

16 Boorse C. On the distinction between disease and illness. *Philos Public Aff* 1975;5:49–68.
17 Boorse C. Wright on functions. *Philos Rev* 1976;85:70–86.
18 Boorse CA. Rebuttal on health. In: Humber JM, Almeder RF (eds). *What is disease?* Totowa, NJ: Humana Press, 1997.
19 Special issue: 'Aristotle, function and mental disorder'. *Philos Psychiatry Psychol* 2000;7.
20 Fulford KWM. *Moral theory and medical practice.* Cambridge: Cambridge University Press, 1989.
21 Hare RM. *The language of morals.* Oxford: Oxford University Press, 1952.
22 Fulford KWM. Teleology without tears: naturalism, neo-naturalism and evaluationism in the analysis of function statements in biology (and a bet on the twenty-first century). *Philos Psychiatry Psychol* 2000;7:77–94.
23 Woodbridge K, and Fulford KWM. *Whose values? A workbook for values-based practice in mental health care.* London: Sainsbury Centre for Mental Health, 2004.
24 Fulford KWM, Williamson T and Woodbridge K. Values-Added Practice (Values-Awareness Workshop), *Mental Health Today*, October 2002, pp25–27.
25 Fulford KWM, Dickenson D and Murray TH (eds). *Healthcare ethics and human values: an introductory text with readings and case studies.* Malden, UK: Blackwell Publishers, 2002.
26 Colombo A, Bendelow G, Fulford KWM *et al.* Evaluating the influence of implicit models of mental disorder on processes of shared decision making within community-based multi-disciplinary teams. *Soc Sci Med* 2003;56:1557–70.
27 Dickenson D and Fulford KWM. Thinking skills: ethical reasoning and problem solving in psychiatric ethics. In: *In two minds: a casebook of psychiatric ethics.* Oxford: Oxford University Press, 2000.
28 Fulford KWM and Bloch S. Psychiatric ethics: codes, concepts, and clinical practice skills. In: Gelder M, Lopez-Ibor JJ, Andreasen N (eds). *New Oxford textbook of psychiatry.* Oxford: Oxford University Press, 2000.
29 Hope T, Fulford KWM and Yates A. *The Oxford Practice Skills Course: Ethics, Law and Communication Skills in Health Care Education.* Oxford: The Oxford University Press, 1996.
30 NIMHE Values website address: www.nimhe.org.uk Click on NIMHE, then click on the NIMHE Values Framework, and download Values Framework.
31 Department of Health. *The ten essential shared capabilities: A framework for the whole of the mental health workforce* (40339). London: The Sainsbury Centre for Mental Health, the National Health Service University, and the National Institute for Mental Health England, 2004.
32 Department of Health. *Delivering race equality in mental health care: an action plan for reform inside and outside services.* London: Department of Health, 2005.
33 Department of Health. *Mental health policy implementation guide: community development workers for black and minority ethnic communities.* London: Department of Health, 2004.

34 Department of Health. *Mental Health Care Group Workforce Team: National mental health workforce strategy* (40276). London: National Institute for Mental Health England; Changing Workforce Programme; Trent Workforce Development Confederation and Social Care, 2004.
35 Department of Health. *National Service Framework for Mental Health – Modern standards and service models.* London: DH, 1999.
36 Department of Health. *Improvement, expansion and reform – the next 3 years: priorities and planning framework 2003–2006.* London: DH, 2002.
37 Welsh Assembly Government. *Healthcare standards for Wales: Making the connections designed for life,* 2005.
38 Nagel T. *The view from nowhere.* Oxford: Oxford University Press, 1986.
39 Sackett DL, Straus SE, Scott Richardson W *et al. Evidence-based medicine: How to practice and teach EBM* (2nd Edition). Edinburgh and London: Churchill Livingstone, 2000.
40 Putnam H. *The collapse of the fact/value dichotomy and other essays.* London: Harvard University Press, 2002.
41 Williams B. *Ethics and the limits of philosophy.* London: Fontana, 1985.

6

Morality and the social brain

ROBIN DUNBAR

Professor Dunbar analyses the evidence for the evolution of the social brain. Drawing from work on primates, he postulates that the sophisticated social strategies observed are possible only because of the relatively large frontal neocortex. The relative size of the neocortex is also related to length of gestation and length of period between weaning and reproduction, which correlates to social learning, the programming of the 'social computer'. He explains the nature of theory of mind and argues that only humans achieve moral behaviour because of the unique computing power of the social brain.

Introduction

Primates are above all social animals: that is their big evolutionary breakthrough. It is what has made them as successful as they have been and, by extension of course, it is what makes humans so successful – we have inherited the same social expertise. What marks primates (or at least monkeys and apes) out as different from all other species of animals is the sheer intensity of their social interactions. The difference between the rest of our primate cousins and humans is simply that we have taken this trend to a whole different level and that is the focus of this chapter. It explores why we are so much better than primates at what all primates do and why, as a consequence, we engage in moral discussions and primates do not (aside, of course, from the observation that they do not have language).

The following example illustrates the kind of behaviour that makes monkeys and apes unusually social by animal standards. The Swiss zoologist Hans Kummer[1] spent many years studying hamadryas baboons in Ethiopia. Hamadryas baboons live in harem-like family units (a male with one to five females), with ten or fifteen of these family units making up a band that lives and stays together. The males are fiercely protective of their females, and will not tolerate them getting near to other males. They do this by punishing the females if they stray too far from them, and particularly if the female allows another male to get between her and the harem male. Kummer once watched what has become one of the classic accounts of a phenomenon known in the animal cognition literature as 'tactical deception'. He described watching a female spend 20 minutes inching her way from where the rest of her family unit was feeding to get behind a big rock. Behind the rock there was a young male from a neighbouring unit and she started to groom with him. It seemed to Kummer that, while the female was behind the rock grooming this young male, she made a very concerted effort to make sure that her head was always visible to her male above the rock as he continued feeding some metres away.

If the male baboon had seen what was going on, there is absolutely no doubt that the female would have been attacked and severely bitten by him. There are two possible interpretations of her behaviour. From a strictly behaviourist point of view, you might argue that she was worried about the consequences of her action, having learned that not keeping within her male's view invited trouble. A more generous cognitive interpretation is that she is thinking something like the following: 'As long as the old man can see my head he will think I am just sitting here innocently behind a rock and so I can get away with whatever it is I am trying to do.' The suggestion on the latter interpretation is that she is manipulating the mental state of her male.

Of course, what is actually going on probably falls somewhere between these two extremes. Perhaps what she is actually doing is not quite so sophisticated as the second interpretation (although such claims have become quite common among scientists who have studied animal behaviour and cognition in recent years). However, irrespective of which explanation is right, behaviour of this subtlety is far from unusual among monkeys and apes – but almost unheard of among any other non-primate species. In the study of animal (and human developmental) cognition, the phenomenon is now referred to as 'mentalising' – that is, being able to understand the minds of other individuals rather than simply working in terms of simple descriptions of their behaviour. The belief is that whereas all other animals function like behaviourists have always supposed (they learn rules of behaviour), monkeys and apes have shifted gear just enough to be able to work in terms of understanding at least a little bit of the mind behind the behaviour.

Despite this upbeat view of primate behaviour, I nonetheless want to argue that there is a limit to which most species of non-human primates can do such things. Indeed, there is even a natural limit to which humans can do this kind of thing, but our limit is much higher than that of our fellow primates and it is this that sets humans apart from our cousins.

The social brain hypothesis

It will be useful to return to the proper starting point for this problem. In the 1970s, the neurobiologist Harry Jerison[2] pointed out that if you plot brain volume for different species of vertebrates against body mass, you end up with a set of relationships that have the appearance of a set of stacked parallel lines. Most important among these is the fact that primates have much bigger brains per unit body mass than all other species of animals. The question this raises is: why should primates have so much bigger brains than other animals? Why do they need what is essentially extra computing power to solve the problems of behaviour that all other animals apparently manage to solve with considerable less computing power? Conventional wisdom dictates that the answer has to do with social evolution: primates need bigger brains than carnivores, for example, because they live in much more socially complex societies and this level of social sophistication is computationally very demanding. This explanation is usually referred to as the 'social brain hypothesis'.

The key evidence for this hypothesis is that if, for a set of primate species, some measure of social complexity such as mean group size is taken (and this is obviously a very simple one) and plotted against some measure of relative brain size (the volume of the neocortex is now normally used) it results in a surprisingly neat straight line. Species that live in large groups have relatively large neocortices. (The neocortex is only considered in these analyses because the neocortex is really the big primate innovation – or at least, to be more precise, exceptionally large neocortices are the primate invention.) The fact that ecological indices produce no coherent relationships when plotted against neocortex volume lends credence to the idea that brain evolution in primates has been driven not by the demands of ecological problem solving (as had been previously assumed) but rather by the demands of social problem solving – the need to bond together large groups of animals into unusually intense relationships based on deep social knowledge of each other.

Humans and the social brain

This finding raises an obvious question: how do we humans fit into this general primate-wide relationship between group size and neocortex size? Since we know what our neocortex size is, it is simply a matter (in principle at least) of plugging the human value into the primate relationship. Doing so gives a predicted group size of about 150 people. I suppose the obvious reaction must be: it cannot be right – humans, after all, go round in very large groupings like nation-states. However, when we are talking about primate groups we are talking about something very much more intimate than what goes on at the nation-state level. The members of the average monkey group see each other every day – they interact on a regular basis, they know each other, they have an ongoing picture in their minds of where everybody else in the group slots in and they use that information on a daily basis to work out just how they should behave. We humans do not even come close to that kind of intimacy in modest-sized towns, never mind nation-states. What we need to identify is the human equivalent of a primate group, and the short answer to that is that we have no idea what it might be.

One solution to this dilemma is to run the problem backwards and ask whether there is any kind of grouping characteristic of humans that is of about the right size. It turns out that there is indeed such a group. If we look at the ethnological data on hunter-gatherers and traditional horticulturalists, we find a regular grouping that is almost exactly 150 in size. What is very characteristic of these groupings is that they invariably have an important ritual rather than a simple ecological function. A classic example is the clans of the Australian Aboriginals. The clan is a group of individuals of more or less fixed membership (aside, obviously, from changes introduced by the usual births, deaths and marriages) who come together once every year or so for a 'corroboree', during which they catch up on news of each other, find out what has happened to Aunty Joan, engage in puberty rituals, arrange marriages and so on. All these are intensely social events: they happen only in the people's minds and we can see no physical instantiation of these groups (other than during the brief period once in a long while when they actually gather together). The rest of the time we would be entirely ignorant of their existence were we not privileged enough to peer inside their minds.

So there is some evidence here that we really do have groups of 150 people. And once primed with that knowledge, it turns out that groups of this kind are exceedingly common in human societies. One example is an informal rule in business organisation to the effect that if you have more than about 150–200 individuals in your organisation, you must have a line-management system of some kind in place: if you don't, information won't get passed on to where it should go and rivalries build up to disrupt the natural processes of interactions between sections. Another example is the number of people in the households to whom we send Christmas cards: this turns out to be very close to 150 on average.

What creates the constraint?

There is, however, a key question we need to ask. Is this apparent cognitive limit on the size of human groups a reflection of a memory overload problem (we can remember only 150 individuals, or keep track of all the relationships involved in a community of only 150) or is the problem a more subtle one – perhaps something to do with an information constraint on the quality of the relationships involved. Two pieces of evidence point to the second as the more likely.

One of these derives from the fact that it is extremely common in primates for there to be a relationship between a male's dominant rank and the number of females he is able to mate. One prediction we can make in light of the social brain model is that the correlation should be much poorer in those species which have a relatively larger neocortex because they can use their big computers to find ways round simple dominance-based strategies. Hence we should find a negative correlation between neocortex volume, on the one hand, and the correlation between male rank and mating success, on the other hand. And this is exactly what we see in the data for monkeys and apes. Those species with relatively large neocortices are able to undermine the dominance of high-ranking males and get the females to mate with them. They do this by exploiting more subtle social strategies – including forming coalitions with other males to undermine the power-based ranks of dominant males, exploiting female preferences.

The second example comes from an analysis carried out by Dick Byrne of St Andrews University.[3,4] He and his colleague Andy Whiten put together an extensive catalogue of examples of tactical deception from the literature on primates. Tactical deception is the term used to refer to cases in which one animal exploits another to gain an objective. One of the classic examples is the case of the female hamadryas baboon deceiving her male while she groomed with another described earlier. Byrne showed that the relative frequency with which these kinds of behaviour are reported for different species correlates positively with the species' neocortex size: species such as chimpanzees and baboons, which have relatively large neocortices, also exhibit tactical deception frequently.

Evidence of this kind pushes us towards the view that it is something about the quality of the relationships that is important, not just their absolute number. We find an upper limit on group size because this is the limit of the number of relationships that an animal can maintain at this level of complexity. It's not just a matter of remembering who is who, or how X relates to Y and how they both relate to me, but rather how I can use my knowledge of the individuals involved to manage those relationships when I need to call on them. We are inexorably pushed towards cognition.

The role of cognition

Narrowing down the options

Some additional reasons for considering the higher cognitive functions to be of overriding importance in this respect come from the anatomical evidence. We looked originally at the whole neocortex, but we can narrow this down a little further by taking out V1, the primary visual area. Doing so sharpens the relationship between social group size and relative cortical volume, suggesting that the key issue has something to do with some aspect of brain function higher up the system. This conclusion is reinforced by an analysis of data on frontal lobe size for the very small sample of species for which these data are available: frontal lobe volume appears to correlate even better with group size. This seems to be suggesting that the key issue is the amount of processing capacity devoted to executive functions in the frontal lobe rather than to the more generalised perceptual and association areas further back in the cortex.

This suggestion is reinforced by the fact that the same analysis for other non-cortical brain areas indicates no such relationship. One obvious possibility in this respect is the amygdala: since the limbic system is implicated in emotional cue processing and emotional responses, we may have anticipated that this sub-cortical region might play a role in social processes. In fact, the amygdala as a whole is unrelated to social group size. However, on a finer scale it turns out that one component of the amygdala (the basolateral complex) does correlate with group size, and what makes this particularly interesting is that this is the one part of the amygdala that has a direct input into the frontal lobe – suggesting once again that executive function plays a key role.

The pre-eminent role that executive function seems to play in this story is reinforced by evidence that socialisation is involved in a significant way. It is generally the case in mammals (and in primates in particular) that total brain size is correlated very tightly with length of gestation (or length of gestation plus lactation), suggesting that brain growth is determined by the length of time for which there is direct maternal investment. However, it seems that what may be true for the brain as a whole may not be true for all its parts. It turns out that, in anthropoid primates at least, the best predictor of relative non-striate neocortex volume (ie neocortex volume excluding the primary visual area) is the length of the juvenile period (the period between weaning and first reproduction). Since this is the period of socialisation, it suggests that learning (perhaps conceived as 'software-programming') plays a crucial role in mediating the relationship between group size and brain size. A second piece of evidence to support this suggestion is that the amount of social (as opposed to solitary or object) play exhibited by different primate species is correlated with relative neocortex size: the more social species are those with larger neocortices.

Possible cognitive mechanisms

The suggestion that executive function and social skills may be important in mediating group size in primates raises the obvious question as to what cognitive mechanisms are involved. The short answer is that we do not really know, but there are some aspects of cognition that seem likely candidates, the most important of which is the phenomenon known in the child development literature as 'theory of mind'. Theory of mind is the ability to understand (or at least imagine) the mental states of other individuals. It is sometimes known as second order intentionality because the individual has to hold two mental states in mind at the same time: he or she has to believe that someone else supposes something is the case. ('Intentionality' is the term used in philosophy of mind to refer to those states of mind that involve beliefs, desires and intentions.) Theory of mind seems to be crucial to much of what humans do, since it underpins much of our social and, indeed, cultural life. In fact, it is crucial for the full development of language, since our ability to exchange information during conversations depends on the hearer's ability to understand just what it is that the speaker is trying (ie intending) to convey by his/her utterances. Language as we use it in everyday situations (as opposed to its written forms) is opaque and telegraphic: we do not say everything we mean but leave it to the hearer to work out what we are trying to say – sometimes deliberately so, in the hope that the hearer actually misconstrues what we say.

Theory of mind is not something that humans are born with: children develop it between the ages of 4 and 5 years.[5] Before this age, children cannot distinguish between their own beliefs about the world and the beliefs of other individuals. And some individuals never develop it: these are the individuals we refer to as autistic. Lack of theory of mind is now recognised as one of the defining clinical features of autistic spectrum disorders.

Developmental psychologists have spent the past two decades studying the development of theory of mind in very considerable detail. The widely accepted benchmark for theory of mind is what is known as the 'false belief' task. This is a task that can be solved correctly only if you understand that someone else is capable of holding a belief that you know (or at least believe) to be false. There are a number of widely used assays for this, including – among many others – the Sally-Ann task and the Smartie task. In the Smartie task, for example, you present a child with a tube of Smarties and ask the child what is inside. The child will typically say 'Smarties'. You take the top off and show that it actually contains pencils. Then you ask the child: I am now going to ask your best friend Jim what is inside... what do you think he will say? Up to the age of 4 years, children are likely to answer 'Pencils' (because they know that this is what the tube actually contains), but after the age of 5 years they are more likely to say 'Smarties'. They are capable of understanding that someone else can hold a false belief.

A role for theory of mind

Although we know a great deal about the development of theory of mind in young children, we hardly know anything at all about its wider natural history. For example, we do not know who else beside humans has theory of mind. More importantly, perhaps, given that theory of mind (ie second-order intentionality) is part of an infinitely recursive hierarchy, we do not really know what the upper limits are for normal adult humans. Understanding the natural distribution of this phenomenon will give us a clearer idea of what the differences actually are between us and other species. So far only two studies have addressed this issue, with the following results.

First, tests of theory of mind in non-human species are confined to just two species, chimpanzees and dolphins. The results for dolphins are fairly unequivocal: they cannot pass false-belief tasks. The results for chimpanzees on the other hand are more ambivalent. Of the two studies carried out so far, one yielded negative results (on a test that young children passed quite readily) and the other yielded modestly positive results. I say modestly here because in no sense could the chimpanzees in this study be said to have passed with glowing colours. However, they did do about as well on this task as children who were just on the threshold of acquiring theory of mind (ie children aged about 4 years old).[6]

The tasks that were presented to the chimpanzees in these two studies were, however, perhaps more challenging than even developmental psychologists might have wished. They involved the chimpanzees having to work out that a human experimenter had a false belief (ie a cross-species inference). In contrast, the standard false-belief tasks used with human children use a human stooge as the agent of the false belief. Might it be that one reason why the chimpanzees performed rather poorly on these tasks was that the tasks were too demanding for them? Perhaps a more chimpocentric task might prove more successful. Follow-up studies in which chimpanzees are pitted against each other do indeed suggest that they may at least be able to understand each other's states of knowledge about the world. In this case, the tasks were not strictly false-belief tasks but rather tasks about the relationship between seeing and knowing ('If another individual cannot see something, does that mean it does not know about it?') so it is not entirely clear how they compare with the standard findings from child-development studies. But perhaps the bottom line is that chimpanzees (and perhaps other great apes) stand just on the threshold of acquiring theory of mind.

The higher levels of intentionality

But even accepting this, it is already clear that chimpanzees are not even in the same league as 6-year-old human children. Human children experience no difficulty at all with the kinds of false-belief task that chimpanzees (and 4-year-old children) struggle to complete. And of course, 6-year-olds are no match for adults. So just how well do normal adult humans do on these kinds of task? Theory of mind, as already noted, is one element in an infinite recursion. Theory of mind is second-order intentionality: I believe that you believe something. In principle, however, we can go on forever in this respect: I believe that you suppose that James thinks that Alice wants Penelope to intend that Edward understands that Peter wants ... Just where do normal adults grind to a halt?

We have carried out what are still the only studies to date that have attempted to answer this question. We have given normal adults short stories detailing a particular social situation and how the various protagonists felt about the situation, and then asked them detailed questions about who thought what about whom. It seems that human adults can happily cope with levels of intentionality up to and including fifth order (I believe that you suppose that James thinks that Alice wants James to suppose [something]) but at sixth-order tasks they collapse. With the exception of a handful of individuals, fourth-order tasks seem to be about our natural limit.

This result has some interesting and relevant consequences in terms of culture. Think of it in terms of literature. The author of a typical novel that involves relationships between three different characters (the minimum to make an interesting story) has to intend that the reader *supposes* that character A *believes* that character B *wants* character C to *think* that something is the case. The writer thus has to achieve fifth-order intentionality. But the reader has to manage only fourth order because he or she does not have to go to the extra level that the writer needs. The fact that fifth order marks the point at which most individuals start to fail these tasks no doubt explains why reading novels (which requires only fourth order) is a widespread phenomenon but writing novels (or, at least, writing novels that others find sufficiently convincing to be worth reading) is extremely rare.

Implications for morality

I want to suggest at this point that the same logic applies to other areas of culture such as morality and, more importantly, the religious systems that underpin morality. If morality is simply a reflection of empathy (and/or sympathy) then it seems unlikely that we really need a great deal more than second-order intentionality: it is only necessary that I understand that you feel something (or that you believe something to be the case). But morality based on this as a founding principle will always be unstable: it is susceptible to the risk that you and I differ in what we consider to be acceptable behaviour. I may think there is nothing wrong with stealing and be unable to empathise with your distraught feelings on finding that I have robbed you of your most treasured possessions. It is not that I do not recognise that you are distraught (or understand what it means for me to feel the same way); it is just that I happen to believe that theft is perfectly alright and that you are making a big fuss about nothing. If you want to steal from me, that is just fine ... help yourself. I may try to defend my possessions, but my view of the world is that possession is nine-tenths of the law and may the best man win.

Giving force to morality

If we want morality to stick, we have to have some higher force to justify it. That does not necessarily mean some kind of religious system, of course. The arm of the civil law will do just as well as a mechanism for enforcing the collective will. But equally, so will a higher moral principle – in other words, belief in a sacrosanct philosophical principle or a belief in a higher religious authority (such as God). The latter is particularly interesting because, if we unpack its cognitive structure, it seems likely to be very demanding of our intentionality abilities. For a religious system to have any kind of force, I have to believe that you suppose that there is a higher being who understands that you and I wish something will happen (such as the divinity's intervention on our behalf). It seems that we need at least fourth order to make the system fly. And that probably means that someone with fifth order is needed to think through all the ramifications to set the thing up in the first place. In other words, religion (and hence moral systems as we understand them) is dependent on social cognitive abilities that lie at the very limits of what humans can manage.[7]

The significance of this becomes apparent if we go back to the differences in social cognition between monkeys, apes and humans and relate these to the neuroanatomical differences between us. First, note that while humans can achieve fifth-order intentionality, and apes can just about manage level two, everyone is agreed that monkeys are stuck very firmly at level one (they cannot imagine that the world could ever be different from how they actually experience it). Their minds are, as it were, thrust so tightly up against the window of experience that they cannot back off far enough to ask whether it could ever be otherwise. Not being able to ask that question means that they can never know what it is to have a false belief and, by extension, they can never imagine a world beyond the world they see: they could never imagine, for example, that there might be a parallel world peopled by gods and spirits whom we do not actually see but who know how we feel and can interfere in our world.

More insights from neuroanatomy

Here, an important piece of the neuroanatomical jigsaw comes into play. If you plot the volume of the striate cortex (the primary visual area in the brain) against the rest of the neocortex for all primates (including humans), you find that the relationship between these two components is not linear: it begins to tail off at about the brain size of great apes. Great apes and humans have less striate cortex than might be expected for their brain size. This may be because after a certain point, adding more visual cortex does not necessarily add significantly to the first layer of visual processing (which mostly deals with pattern recognition). Instead, as brain volume (or at least neocortex volume) continues to expand, more neurones become available for those areas anterior to striate cortex (ie those areas that are involved in attaching meaning to the patterns picked out in the earlier stages of visual processing). An important part of that is, of course, the high-level executive functions that are associated with the frontal lobes. Since the brain has, in effect, evolved from back to front (ie brain size increases during primate evolution are disproportionately associated with expansion of the frontal and temporal lobes), it is precisely those areas associated with advanced social cognitive functions that become disproportionately available once primate brain size passes beyond the size of great apes. Indeed, great ape brain size seems to lie on a critical neuroanatomical threshold in this respect: it marks the point where non-striate cortex (and especially frontal cortex) starts to become disproportionately available.

It seems to be no accident that this is precisely the point at which advanced social cognition (ie theory of mind) is first seen in non-human animals. Moreover, when achieved levels of intentionality for monkeys, apes and humans are plotted against frontal lobe volume, we get a completely straight line. That, too, seems to be no accident.

Conclusion

So, we have arrived at a point where we can begin to understand why humans – and only humans – are capable of making moral judgements. The essence of the argument is that the dramatic increase in neocortex size that we see in modern humans reflects the need to evolve much larger groups than are characteristic of other primates (either to cope with higher levels of predation or to facilitate a more nomadic lifestyle). After a certain point, however, the computing power that a large neocortex brought to bear on processing and manipulating information about the (mainly social) world passed through a critical threshold that allowed the individual to reflect back on its own mind. Great apes probably lie just at that critical threshold. With more computing power still, this process could become truly reflexive, allowing an individual to work recursively through layers of relationships at either the dyadic levels (I believe that you intend that I should suppose that you want to do something ...) or between individuals (I believe that you intend that James thinks that Andrew wants ...). At that point, and only at that point, religion and their associated moral systems can come into being. In terms of frontal lobe volume expansion, the evidence from the human fossil record suggests that that point is likely to have been quite late in human history. It is almost certainly associated with the appearance of the human species around half a million years ago.

References

1. Kummer H. *In quest of the sacred baboon*. Princeton: Princeton University Press, 1997.
2. Jerison H. *Evolution of the brain and intelligence*. New York: Academic Press, 1973.
3. Byrne RB. *The thinking ape*. Oxford: Oxford University Press, 1995.
4. Byrne RB, Whiten A (eds). *Machiavellian intelligence*. Oxford: Oxford University Press, 1988.
5. Ashington JW. *The child's discovery of the mind*. Cambridge: Cambridge University Press, 1994.
6. O'Connell S, Dunbar RIM. A test for comprehension of false belief in chimpanzees. *Evol Cognition* 2003;9:131–9.
7. Dunbar RIM. *The human story*. London: Faber, 2004.

Part 2
THE SOCIAL SCIENCE

7

Demons and angels: who cares?
Poor care structures and moral breakdown

CAMILA BATMANGHELIDJH

Camila Batmanghelidjh relates her experiences in dealing with children and young people in a deprived area of London. Most of the children were conditioned by violent, abusive upbringings where authority derived from physical power. Batmanghelidjh shows how the opposite environment, of physical and emotional support, leads to a self-regulating group of young adults. She illuminates the connection between these environments and the children's behaviour, body language and emotional poverty. Failure to grow socially is equated with acquired failure to empathise with others.

In the past few years the pages of our newspapers have been filled with descriptions of children as 'demonic and dangerous' – young people drug dealing, carrying out street robberies and murdering. Middle England despairs and barricades itself behind remote control gates hoping to keep out this 'other world'. It is a drama of 'angels' and 'demons', victims and perpetrators, neither really understanding their contribution to the epic.

In this chapter I hope to illustrate how the 'angels' – decision makers, comfort seekers and those who pledge to repair – share responsibility for the breakdown of morality with the children they accuse.

Kids Company: working with vulnerable children and young people

My introduction to the problem began approximately ten years ago when I first set up Kids Company, a charity offering support to exceptionally vulnerable children.* Trained as a psychotherapist in the privileged schools of

---

*Kids Company (www.kidsco.org.uk) was founded two years after my first charity, the Place 2Be (see www.theplace2be.org.uk), see *Shattered lives: children who live with courage and dignity.*[1]

Hampstead, I ventured across London to Peckham intending to provide a holiday programme for vulnerable children who needed care when schools were closed. A local minicab company led us to some disused railway arches, which we converted into a welcoming space for young children. The word of a new club opening had spread on the streets, and within the first week of opening we were greeted by approximately 100 adolescent boys. They came to give us a challenge!

My early childhood in Iran had been in the lap of luxury, and I then believed that every child who was born was at risk of being kidnapped and needed two police bodyguards. My protection I attributed to my being a child and not to my father's enormous wealth. My background and the therapy training schools of Hampstead had done little to prepare me for the client group I had to face at 3pm every day we opened.

The boys were an enigma; I did not understand a word they said as their street language operated beyond the thesaurus. I had no idea 'bad' meant good and 'bling bling' was gold. The boys entertained themselves by pulling out knives and ripping the furniture, using their cigarettes to burn holes into the sofa. Whatever they could steal, they would. They rolled their joints and smoked themselves into higher levels of aggression. I felt completely out of my depth and was acutely aware of the potential for a riot. The team I had with me felt terrorised and de-skilled. Upon opening most of them would dive behind the doors for safety. Despite the meeting of two sets of complete cultural strangers, the boys came back every day and I held on to my Persian hospitality, courteously welcoming them and urging them not to spit or burn the place down! It was my sense of vocation and morality which sustained me through this breathtakingly difficult time. Even in remembering, the blood drains out of my cheeks; I was terrified of my own uselessness.

The boys now tell me that they found my gentleness a surprise and they expected me to give up, which I never did. In order to talk to them and understand their experiences, very early on we employed some adult males from the community. They came from similar backgrounds to the children and acted as translators for us. Very soon I learnt the street business and slang, managing to keep up with the pace of it. The boys began trusting and wanting to share their life stories. Kids Company as an organisation evolved because of what these young people shared with us.

Within a year we had some 300 children of all ages making their way to our service, having heard about it on the street. These were some of the most vulnerable and traumatised children whose life stories make shocking reading.

Through trial and error we developed a service which met many of their unmet needs. This included three meals a day, often a lifeline to children of drug addicts who were being left without food. We set up a school on site because so many young people had been out of education and were unable to access it, through a combination of personal shame and the school's unwillingness to have them back. Our social workers visited the children's homes to find shocking conditions of poverty and neglect: children without beds or bedding; one household of eight sharing three plates at meal times, with no chairs to sit on; a toddler running to stuff the rat holes with rags so that the rats would not bite the baby. We took local authorities to court for their failure to honour their statutory duty of housing children, and we won every case, building up a credit rating amongst the kids and a hate rating amongst the officials.

Our staff took children with their rotten teeth to the dentist and girls involved in prostitution to sexual health clinics. Where children were not registered with GPs we begged for their registration. Over time we acquired a reputation for meeting children's needs, and clients whose problems we had solved brought their peers, some desperate to survive, others tired of living.

A powerful community of children seemed to have emerged in the neighbourhood. The older boys and girls survived through crime, and the younger ones failed to thrive and met their basic needs by shoplifting for food. What all these children shared in common was the absence of a competent carer in their lives. Many lived with worry for their parents' well-being and struggled to shield their younger siblings as best they could. Every day I observed their desperation, their depletion and their remarkable courage. As they shared their stories with me, I came face to face with the so-called 'demons'.

The media zoomed its lenses into Peckham to cover the murder of Damilola Taylor, a young and innocent boy of ten, murdered mindlessly at the hands of teenagers. He was perceived as the 'angel' and his killers as the 'demons'. The police woke up to an underground of children who all knew each other and fiercely protected their world from the intrusion of adults. The investigators were shocked at the children's lack of cooperation and their active misleading of the investigators. Middle England woke up to a world they came across only when they had the misfortune of being its victim. As the young people were dragged through a trial, I was amazed at how no adult or carer was held accountable. The children accused of that crime had been victims themselves for many years; their emotional life had been killed off well before they were suspects in a murder trial.

Unlike the public, we knew them and their histories – a catalogue of early childhood neglect and abuse, both within their homes and within the institutions which had pledged to care for them. The children taught us an important lesson: the making of a 'demon' starts with the failure of the 'angels'.

Impoverished upbringings and damaged psyches

The 400 life stories we have now accumulated point to a rotten care structure that fails to nurture and protect children. The psychological journey is one of a child victim who is terrorised and resorts to emotional perversion in order to survive. Sometimes they become perpetrators. Universally the children described being born into households with high levels of chaos and danger. One young boy slept with a knife under his pillow terrified of the drug dealers who burst into his house. His mother supported her drug habit through prostitution; his stepfather facilitated it. Before the age of 16, the boy had stabbed his stepfather twice, resulting in hospitalisation. He was never held accountable, as the man knew the boy had risen in defence of his mother who was being battered.

Young children playing in our dolls house would pick up toy guns and act out the shootings in their homes. Mothers and children were separated as the women were imprisoned for 5–7 years for drug trafficking, leaving toddlers in the care of teenage brothers who could barely manage their own lives. Exhausted teenage mothers gave birth to babies, unprepared for the sophisticated task of parenting. Early attachment disturbances rupture the child's need for a positive reflection of self. Infancies disturbed by violence and mindlessness result in toddlers greeting the world with an impaired sense of security and agency.

Initially these traumatised children are victims, passively begging for change. They plead, sometimes they protest, and when they realise there is no hope for change they undergo, as we see it, a psychological transformation. In a depleted but determined way they shut down their capacities to feel. Numb to the world, their environmental disturbances cease to upset. In the short term this is a useful strategy and an instinctive reaction to protect against trauma. Sustained over a long period of time the numbing of emotions results in an impoverished emotional life and therefore poor moral comprehension. Morality cannot be taught without a meaningful emotional context; it needs to be experienced to be believed. These emotionally cold children cease to exercise their emotionality in reciprocal exchanges. They elicit no tenderness from others and soon cease to access it within themselves. Protected against pain, they forget what sorrow feels like; they cannot feel empathy because they cannot access hurt. In not being able to feel for themselves they are unable by proxy to feel for their victims. Their inability to remember pain separates them from remorse. Their lack of guilt paralyses their ability to make genuine reparation. Years of being emotionally cold reflects back at them a self which like a rubber doll is void of feeling. As their distorted sense of self is not altered by an alternative perception or experience, they believe all those they come into contact with to be equally feelingless. The mindset of 'dog eat dog' and 'mind your own business and survive no matter what the cost' becomes a world view in which others are perceived to be the same.

Sometimes these children are capable of great cruelty. Their early brutalisation leads to a breakdown of prohibitions against violence. When their victims beg for mercy, these children hate the pleading as it reminds them of their own victimisation. They often show no mercy as they were afforded none.

These emotionally exhausted young people are brave, as they do not fear death. They walk around spiritually and emotionally ravaged; threats of deprivation, imprisonment and physical death are viewed sarcastically. We know from research in the field of neurobiology that this emotional shutdown is paralleled on a physiological level. Infants subjected to poor-quality care, disorganised attachments and trauma are exposed to levels of fright, leading to atrophy of key neuronal pathways affecting pro-social behaviours and memory centres.[2]

Chronically distressed children live with either overcharged or exhausted adrenal functioning. This is the young boy or girl whose hyper-aroused state seeks to be reflected back in an externally charged event, either an environmental trauma which they have to dodge or one which they have to create in order to have their sense of inner urgency reflected back at them, affirming their existence. The exhausted child is too depleted to seek fight or flight; he or she ceases to care even to self-preserve and passively waits to die, sometimes walking in the path of an accident or a fight. Both these sets of children often turn to drugs in order to self-tranquillise or manically escape themselves.

These young people are sensitive to humiliation; they have been exposed to a catastrophic loss of power and often their humiliation has been confirmed in the onlooker's gaze. A profound experience of their worthlessness makes them exceptionally sensitive to any encounter which may risk their dignity. Sometimes they interrupt the onlooker's gaze for fear that the observer may perceive them as flawed. There is no tolerance for being viewed. Other people's looks are anticipated as generating violence and shame. In a hyper-aroused state, insults and facial expressions are all deemed as a potential for annihilation. To avoid them, the child being 'looked at' ceases to be passive and takes charge in an attack. Prematurely they may assault the onlooker or rebuff them by an aggressive 'What are you looking at?' Humiliation is about a profound sense of disgust which is attributed to the self. The humiliated person feels intrinsically flawed and wishes not to be reminded. Those who feel shame regret a negative act, which in some way has led to a lowering of an otherwise intact self-esteem.

One way in which these humiliated children often avoid their sense of powerlessness is by wearing and acquiring designer goods. Designer wear on the street signals an individual who has been criminally competent enough to meet his or her own needs. This individual by definition is not likely to be a powerless victim and therefore should not be attacked or challenged. The designer goods are believed to restore a lost dignity and act as a portable living room, showing off an illusion of equality to their wealthier counterparts. Many of these young people are too ashamed to take friends back to their depleted homes, so the designer wear affords an escape from humiliation.

Frequently, the young people are helped to acquire material goods by being recruited into the drug trade; adult male dealers run young children as mules, carrying drugs from one dealer to another. On this journey the child learns a perverse trade for which he is rewarded with his own drugs to deal or sell. In poor neighbourhoods the drug economy sustains the population above the poverty line. It turns around a vast amount of money and is left unchallenged because of a shortage of policing. To protect this business, dealers use violence. It is in this context that children learn how to use knives and firearms. Boys cry with terror as they describe their dealer pushing the barrel of a gun down their throat for a minor misdemeanour. Many of these boys do not have a father figure at home and these drug dealers become their masculine role models, perpetuating a cycle of violence and abuse.

Providing practical and emotional support

In order to return these young people to the centre of society as law-abiding citizens whose contributions would be respected, Kids Company needed to intervene on both emotional and practical levels. Our practical strategy involved ensuring the young person's basic needs were being adequately met. We found them somewhere to live, provided them with the basic necessities to run an adequate home, bought them clothes and ensured they had food. Once their physical needs were met, we attempted to introduce a gentle sense of routine.

Experience has taught us that these young people find it difficult to adhere to structures which were initially demanding. Each client was encouraged to attend education on our premises for an hour a day, as long as they arrived between 10 am and 5 pm. Most young people felt proud of being able to achieve this commitment; gradually they became involved in gaining qualifications, accessed partly through us and partly through other education providers. One of our workers ensured that college attendance was maintained by speaking to tutors and assisting in homework assignments.

These practical gains were paralleled by engagement in emotional tasks. Our workers were encouraged to establish an attachment relationship with the young person, often functioning in a parenting capacity. Kids Company workers have been trained to apply therapeutic thinking in a meaningful context relevant to the life experiences of our children. Discussions about drug dealers, guns and drugs took place alongside the need for a bus pass or clean socks. Children felt that we had taken on board the reality of their existence and they did not need to protect us from the rottenness of their lives. The same spirit of honesty permeated the relationship as children roller-coasted through feelings of attachment and terror of betrayal. Destructive and anti-social behaviours were perceived to be in the context of emotional hurt. We rarely used sanctions except for a premeditated disregard of the rules, for example attempting to smoke drugs on site.

As young people were protected and looked after, they would often soften both emotionally and physically. Their iciness gave way to warmth; they would feel grief and sometimes fear that their newly acquired tenderness would prohibit them from being able to fight on the street. Peers reacted in two ways: some joined the young person in eliminating violence from their lives, and others mocked them for 'going soft'. One of our clients was nearly burnt in his hostel. Having been the leader of a Peckham gang, others hoped to acquire his status by destroying him. He was plunged into confusion and for a while went back to the street in order to reclaim his position.

The 'to-ing and fro-ing' of recovery describes the journey on which clients and workers follow in these neighbourhoods. The gains of our clients were always challenged by environmental disturbances. For this group of young people, a short-term intervention is futile. They require long-term consistent input on a profoundly emotional level in order to be able to embrace change. Workers coming into contact with them experience themselves as emotional punch bags, simultaneously hugged and assaulted. There is nowhere to hide; the children demand a truthful presence and they need to be reflected back as individuals for whose survival the worker has a passion. Often these young people can treat the worker as a victim, bullying, terrorising and insulting them with the same level of harshness to which they were subjected as children. This is not a job limited to the confines of an office; it requires risk taking and repeated fresh starts without being punitive or judgemental. It is easy to see how poorly supported workers could despair and give up, believing change to be an impossibility.

Many young people do make a recovery, and have gone on to employment, college and university. Although the outside world measures success in terms of these achievements, the children of Kids Company recognise success differently: in being moved from a position of not caring to one where they can access tenderness and compassion. Their feeling of hopelessness diminished in the presence of hope. It began with our workers hoping for better things on behalf of the child whilst waiting for the child's capacity to aspire to awaken.

Professional and institutional responses

So far I have described the so-called 'demons', but what of the so-called 'angels'? The children bring to the equation a psychology at ease with cruelty, but the 'angels' may also unwittingly contribute to a moral breakdown. As a society we have pledged to protect and educate children; in some areas we do both badly, and excuse it with lack of resources.

Passionate and vocational workers enter our care professions wanting to deliver excellence but within a short period of time they too may become morally exhausted. A worker who is repeatedly exposed to pain and dysfunction which they cannot ameliorate may, like the child, acquire profound feelings of being rendered powerless and humiliated by their ineffectuality. If care agencies contain workers who in order to survive have shut down their capacities to feel, then both these carers and the children live in hopelessness, neither able to inspire the others' emergence. The capacity to hope is a fundamental ingredient of morality; when we make mistakes we live in the hope of reparation and we project ourselves beyond the despair. If hope dies it kills aspirations.

In a culture where career achievements sometimes substitute for personal identity, few professionals are prepared to speak up about their failure to deliver good-quality care. This failure to care encompasses large numbers of children in deprived neighbourhoods throughout London. One borough recorded some 570 children without a school placement. In another, the social work department could only take on cases of sexual and extreme physical abuse, leaving neglect outside its doors.

As these doors shut in the faces of vulnerable children, the brutalisation is sustained and children's perceptions of adults as corrupt and non-caring are reaffirmed. Children who are not extended a helping hand continue to believe themselves alone in the world; they have no concept of a wider society which can be held accountable or which should be honoured. When they needed society to protect them they were betrayed.

Currently, youth offending teams have the greatest power and budget to meet children's needs but sadly a child must acquire a criminal label before becoming eligible. This perception is supported by the fact that in one local authority there were fewer than 200 children on the child protection register and some 1,500 on the youth offending register.

The 'angels' of middle England sometimes find their quality of life affected by the crimes of poor children in their neighbourhoods, and demand an intervention – revenge and punishment – not knowing that often the criminal child is a child in need. Years before, these children were pleading for help behind closed doors, too powerless to impact on other people's quality of life; they were left unheard and they grew up not listening and not caring.

There is a fantasy that punishment will redress the moral balance: children are threatened with prison, anti-social behaviour orders (ASBOs) and tags that limit their movement. Often children are returned and locked for the night in the very households where abuse demanded an escape. Prison may be welcomed, as children are afforded consistency, warmth and a measure of safety sometimes for the first time. Punishment never teaches morality. Disturbed children have disorganised memory structures, avoid noxious stimuli and can seldom appropriately process a punishment to inform their future actions. Instead, they often perceive the punishing adults as hateful and in collusion with all others who have harmed them over the years.

It is not about money: in erecting the Millennium Dome the government was able to spend in one year the equivalent to ten years of this country's child mental health budget. It is about priorities and making a shift from pleasing the voter in the short term to making thoughtful decisions about the long-term emotional well-being of our society. While tenderness and compassion are seen to be expressions of individual choices, morality is a social expression and fundamental to the health of our society.

Conclusions

I hope I have been able to demonstrate in this chapter that depleted communal morality cannot be the fault of 'demons' alone. A morally healthy society is a product of the contributions of all its members. As long as the story is polarised into a discourse about 'angels' and 'demons', we are unlikely to find a solution.

The breakdown of morality has its origins in the breakdown of care, not inadequate surveillance cameras. A lasting solution has to commence with robust financial investment in our care structures and a better understanding of its pathogenesis, supported by the facilities it deserves. The importance of the care professions must be acknowledged and afforded the support they deserve. Within these professions we need to rekindle the sense of vocation, and the striving for excellence, as their strength rests in their spiritual dimension. If our society is seeking moral rehabilitation both 'angels' and 'demons' need guidance and help to fulfil their respective responsibilities and contributions to our social fabric.

The sense of morality is partially a by-product of compassion and care; it cannot be created in a vacuum. Its breakdown points to a breakdown in our care structures. Its repair must be our collective responsibility.

References

1 Batmanghelidjh C. *Shattered lives: children who live with courage and dignity.* London: Jessica Kingsley Publishers, 2006.
2 See for example Schore A. The effects of early relational trauma on right brain development, affect regulation and infant mental health. *Inf Mental Hlth J* 2001:22;201–69.

8

Biological, psychological and social factors in the pathogenesis of psychopathy

MICHAEL L PENN, AMER PHARAON AND ATILLA CIDAM

Professor Penn and his colleagues, Pharaon and Cidam, explain the socio-biological factors in moral development and the aetiology of psychopathy. Antisocial behaviour is related here to an insensitivity to the normal facilitatory and inhibitory stimuli of traditional upbringing. They argue that without the conditioning effect of these stimuli, a sense of justice cannot emerge. Deliberation of right and wrong is compromised to the extent that moral awareness does not emerge. They also suggest that the somatic root of the neurobiological dysfunction seen in psychopathy is a deficit in the autonomic arousal system. The aetiology of this lacuna in moral reasoning is examined in other chapters in this book.

Introduction

This chapter explores the idea that moral development depends on the operation of the laws of causality (or, more specifically, the laws that govern contingency), and that processes which distort the operation of these laws, or which disrupt normal perceptions of them, make it harder to acquire moral qualities. In extreme cases, this difficulty will manifest itself as a constellation of character flaws that can be encompassed in a diagnosis of antisocial personality disorder or psychopathy.

We also suggest that while the *psychosocial* roots of antisocial personality disorder can be traced to experiences with injustice,* the *somatic* roots can be traced to a dysfunction in neurobiological processes necessary for an adequate

---

*Distinctions are sometimes made between 'psychopathy', 'sociopathy' and 'antisocial personality disorder' (APD). The differences between APD and psychopathy hinge upon the

continued over

perception of the mechanisms of reward and punishment. Inasmuch as injustice is a wilful distortion of the principles of causality and contingency, and a proper perception of rewards and punishments is indispensable to acquiring an understanding of justice processes that distort contingency, or that disrupt normal perceptions of it, will be associated with significant disabilities in the acquisition of moral values. In extreme cases, these moral disabilities will manifest themselves in anti-social behaviour or in a constellation of character flaws that are encompassed in a diagnosis of antisocial personality disorder. In other words, we claim that justice is a sine qua non of moral development. Since all forms of moral development involve volitional processes, however, antisocial personality disorder may also develop as a result of chronic, wilful misuse of otherwise healthy human capacities, even in situations where there is no absence of justice. We also review research in the biological, social and philosophical sciences that serve to support and illuminate these claims.

Historical overview of the nature and pathogenesis of psychopathy

The concept of psychopathy emerged early in the 19th century with the work of the French psychiatrist Phillippe Pinel. In an effort to advance a new theory on crime, Pinel abandoned the view that madness necessarily involved an intellectual disturbance and advanced the concept of *'manie sans délire'* (insanity without delirium). Pinel was attempting to characterise individuals he saw in practice who engaged in impulsive and self-destructive acts despite the fact that their ability to reason seemed unimpaired. Regarding such individuals Pinel wrote, 'I was not a little surprised to find many maniacs who at no period gave evidence of any lesion of understanding, but who were under the dominion of instinctive and abstract fury, as if the faculties of affect alone had sustained injury'.[3]

---

methods used for assessing diagnostic criteria. While antisocial personality disorder and sociopathy rest upon assessments of behaviour, diagnoses of psychopathy require assessments of long-standing personality traits. And although the diagnostic criteria that distinguish sociopathy from psychopathy have been difficult to resolve, David Lykken, one of the foremost authorities in the field, has argued that while sociopathy refers to individuals whose unsocialised character is due primarily to parental failure, psychopathic personality disorder (or psychopathy) is due primarily to biological deficits that have had a deleterious impact on early socialisation.[1] Other researchers, such as Black, lump all three terms together, arguing that psychopathy is merely 'the older, more chilling' label for our current identical conceptions of sociopathy and APD.[2] For our purposes here it is sufficient to note that the behavioural or personality dysfunction in the moral domain involved in each of these conditions is significant enough to be indicative of a profound failure of moral development.

Benjamin Rush (1812), the well-known American physician who signed the Declaration of Independence, wrote about similar, perplexing cases where some of his patients manifested lucidity of thought but socially deranged and destructive behaviours. He described these individuals as having an 'innate, preternatural moral depravity' in which 'there is probably an original defective organisation in those parts of the body which are preoccupied by the moral faculties of the mind'.[4] Not long after, the British psychiatrist James Prichard[5] coined the term 'moral insanity', which distinguished between two types of insanity, one affecting the intellect and the other affecting emotions and will. According to Prichard, an individual afflicted with the latter was 'incapable of conducting himself with decency and propriety in the business of life. His wishes and inclinations, his attachments, his likings and disliking have all undergone a morbid change'.

Prichard, like Rush, dissented from Pinel's morally neutral attitude towards the disorder and became the major exponent of the view that it signified a reprehensible defect in character that deserved social condemnation.[5]

Around the same time, the Austrian physician Franz Joseph Gall was formulating the new science of phrenology. According to this now discredited pseudoscience, human behaviour is regulated by 27 different faculties or propensities, each located in a particular part of the brain. The over-development of three of the propensities, greed, self-defence and carnivorous instinct, could lead to the emergence of criminal behaviour. Phrenology, which flourished in the middle of the 19th century, integrated Pinel's and Prichard's theories into its own framework. According to Gall, each mental faculty was localised in a distinct part of the brain, functioning independently in relative isolation from the others. Therefore, one of the brain's faculties (the intellect) could function normally while another (the moral faculty) lay dormant. Using this rationale, phrenologists encouraged the association of moral insanity to brain malfunction. Whereas Prichard and Pinel viewed *manie sans délire* and moral insanity as afflictions with no physical causes or consequences, phrenologists tied these afflictions to brain matter itself. Ultimately, phrenology heavily influenced late 19th-century thinking about criminality as mental illness.[3]

Near the end of the 19th century the term 'moral insanity' began to elicit many strong objections. While religious communities disputed the idea that certain individuals were inherently incapable of doing the Lord's work, the legal community refused to accept that individuals could be deemed insane outside a court of law.[6] In an attempt to quell the rising objections, Koch, a German psychiatrist, turned his attention away from value-laden theories towards an emphasis on observational research. As a result, he replaced the term 'moral insanity' with 'psychopathic inferiority', under which he included 'all mental irregularities, whether congenital or acquired, that influence a man in his personal life and cause him, even in the most favorable cases, not to be fully in possession of normal mental capacity'.[3] This new term was used to describe a range of psychological disorders that today would be called personality disorders. Ultimately, Koch's attempt to explain psychopathic behaviour as the direct result of biological abnormality was undone by emergence of the concept of 'sociopathy'. Partridge, an American psychiatrist, first used the term 'sociopath' as a means of signifying the psychopathic personality's ostensibly social origins.[7]

Cleckley and Hare's criteria for psychopathy

The most incisive, thorough and influential clinical characterisation of the psychopathic personality was provided by Harvey Cleckley in his seminal book, *The mask of sanity*, published in 1941.[8] In this work, Cleckley developed the first coherent description and added a set of diagnostic criteria that could be used in assessing those suspected of suffering from the condition. These criteria are set out in Table 8.1.

In his clinical studies of psychopaths, Cleckley concluded that their personalities lacked the normal affective accompaniments of experience. Simply put, in Cleckley's view, the feelings that normally accompany experience are unusually muted in the psychopath. Furthermore, whereas Prichard and Rush believed that the psychopathic personality involved an innate lack of moral sensibility, Cleckley assumed that moral feelings and compunctions are not God-given but must, at least in part, be learned. However, inasmuch as empirical research had shown that the involvement of emotions is necessary

Table 8.1. Cleckley's criteria for assessing psychopathy.

- Superficial charm and good intelligence
- Absence of delusions and other signs of irrational thinking
- Absence of 'nervousness' or other neurotic manifestations
- Unreliability
- Untruthfulness and insincerity
- Lack of remorse or shame
- Inadequately motivated antisocial behaviour
- Poor judgement and failure to learn from experience
- Pathological egocentricity and incapacity for love
- General poverty in major affective reactions
- Specific loss of insight
- Unresponsiveness in interpersonal relations
- Suicidality rarely carried out
- Impersonal sex life
- Failure to follow any life plan
- Fantastic and uninviting behaviour.

to the learning process, Cleckley reasoned that the socialisation of the psychopath, who appears to be relatively incapable of emotional reactions, would likely prove to be problematic.

Hare has used Cleckley's formulation of the psychopathic personality as a basis for his re-conceptualisation of the disorder (Table 8.2).[9] His Psychopath Checklist (PCL) and its revision (PCL-R) are now universally used by researchers who study the condition. Two sets of interrelated personality factors have emerged from Hare's work on the psychopathic personality. The first set describes a narcissistic personality variant of psychopathy, including tendencies toward selfishness, egocentricity and a lack of empathy, while the second more directly relates to those with an overtly antisocial lifestyle, for example one with impulsivity, early periods of delinquency and a tendency towards parasitic patterns of relating to others. The three final items on the PCL do not correlate strongly with either main set of factors; they are, nevertheless, commonly-identified behavioural characteristics of individuals with the disorder.[9]

Table 8.2. Hare's Psychopath Checklist.

Narcissistic personality	Overtly antisocial lifestyle	Other common behavioural characteristics
Glibness/superficial charm	Need for stimulation/ proneness to boredom	Promiscuous behaviour
Grandiose sense of self-worth	Parasitic lifestyle	Many short-term marital relationships
Pathological lying	Poor behavioural controls	Criminal versatility
Cunning/manipulating	Early behavioural problems	
Lack of remorse or guilt	Lack of realistic, long-term goals	
Shallow affect	Impulsivity	
Callousness, lack of empathy	Irresponsibility	
Failure to accept responsibility for actions	Juvenile delinquency	
	Revocation of conditional release	

Lykken's low fear quotient theory

Recent research on psychopathy has focused on the affective characteristics of the disorder, specifically the deficiencies in emotional responsiveness. Invoking Cleckley's observation that the psychopath lacks the normal affective accompaniments of experience as a basis for his work, Lykken suggested that the psychopath suffers from an attenuated experience, not of all emotional states but specifically of anxiety and fear.[1] This alternative proposal was labelled the 'low fear quotient theory, according to which all individuals can be placed somewhere along a continuum of vulnerability to fear. Individuals on the low end of such a continuum are at high risk of developing psychopathy, while those who are characterised by fearful inhibitions are especially unlikely to develop such a disorder. Thus, in this model, an individual's *invulnerability* to the experience of fear is seen as a potential risk factor in the pathogenesis of psychopathy.

The model proposes that a biological deficit in the autonomic arousal associated with fearfulness serves as a diathesis for the development of sociopathy in childhood. This is because such a deficit renders it more difficult for a child to develop a sense of moral conscience. Indeed, a significant part of a child's acquisition of moral discernment depends upon the negative consequences that normally attend antisocial behaviour. Thus, Lykken argued that children who are especially vulnerable to antisocial personality disorder, however, are less likely to be deterred from antisocial or immoral behaviour because they do not perceive its negative consequences.

Robert Hare designed a widely cited study that tested Lykken's low fear quotient theory.[10] Both psychopathic and non-psychopathic subjects were told in advance that they would experience an aversive stimulus at the end of a countdown. The countdown began at nine and continued to zero at intervals of three seconds. Individuals diagnosed with psychopathy showed relatively little electrodermal arousal compared with non-psychopathic subjects during the countdown. In addition, non-psychopathic subjects showed higher arousal at the start of the countdown with a larger and earlier increase in skin conductance as the count approached zero.[10] The anticipatory arousal that often accompanies the expectation of pain was not as salient in individuals suffering from psychopathic personality disorder.

In the most cited study in this area, Lykken examined the effects of punishment on learning in a group of individuals who were described as 'sociopaths'.[11] Lykken's laboratory-based research consisted of the subjects learning to press a 'correct' lever, and Lykken was interested in finding out which group of subjects most successfully learned to avoid physical punishment.

Subjects sat in front of a panel with four levers. Immediately above each of the levers was a red light and a green light. The subjects' task was to locate and press the lever that turned on the green light in a series of 20 trials. As the correct lever changed on each trial, subjects were required to remember their sequence of responses from the first trial to the one they were now working on. In effect, a complex pattern had to be learned. During each trial, the subject was presented with four choices, only one of which would turn on the green light. Two of the levers turned on a red light – thereby indicating an incorrect answer – while the third delivered an electric shock. Inasmuch as the paradigm provided two kinds of feedback, one that simply indicated that the subject had made the wrong choice and another that delivered a physical punishment, the investigators were able to document the specific kinds of negative experiences that individuals who score high on psychopathy could not learn from. They found that psychopaths have considerable difficulty learning from negative experiences that are generally perceived to be physically painful.

As expected, there were no differences in the total number of mistakes made by subjects who scored high on psychopathy compared with those who did not. However, whereas the people not diagnosed as psychopaths quickly learned to avoid the electrified levers, the people with psychopathy scores made the most errors that led to electric shock. These findings suggest that the latter's particular learning deficiency consisted of an inability to learn from painful experiences. Essentially, Lykken's study suggested that punishment, or threat of punishment, did not seem to influence the psychopathic subjects' behaviour.

Quay likened psychopathic conduct to 'an extreme of stimulation-seeking behaviour'. He theorised that psychopaths possess an abnormality in their physiological reaction to sensory input that leads them to seek higher levels of stimulation. Quay believed that 'much of the impulsivity of the psychopath, his need to create excitement and adventure, his thrill-seeking behaviour, and his inability to tolerate routine and boredom is a manifestation of an inordinate need for increases or changes in their pattern of stimulation.' Because the psychopath finds it extremely difficult to maintain such an optimal level of stimulation, reaching these thresholds brings great pleasure.[12]

In 1977, Quay elaborated upon his physiological theory by suggesting that environmental factors also play a significant role in the development of adult psychopathy. His new formulation suggested that children who have high physiological thresholds are likely to be natural stimulation seekers. The behaviours associated with stimulation seeking are often met by parental hostility, rejection and inconsistent discipline. Children who face such parental reactions are likely to increase their deviance, thereby increasing their parents' frustration and creating a vicious, character-corrupting cycle.

Researchers have attempted to test Quay's theory in two different ways.[13] The first attempts to use several psychometric instruments to measure an individual's degree of sensation seeking. The second method has involved experimental manipulation of stimulation type and intensity, thereby allowing researchers an opportunity to observe differences in reactions between psychopathic and non-psychopathic subjects. Research has found consistently that psychopaths act in ways that appear to be sensation seeking. Fairweather explored the learning rates of psychopaths under three different conditions: certain reward, uncertain reward and no reward. His results indicate that psychopaths learn best when a reward is uncertain.[14] According to Quay, these findings support his theory. He reasoned that psychopaths preferred the reward to be uncertain, and learned better under such conditions, because the uncertainty heightened arousal and thereby facilitated learning.

Similarly, Hare tested the learning rates and the desire for stimulation in psychopathic and non-psychopathic subjects. Knowing that given a choice most subjects prefer to receive an electric shock immediately rather than after a ten-second delay, Hare gave his subjects a choice on each of six trials between immediate or delayed shock. Past research has found that the delayed shock was viewed as more distressing and, hence, subjectively more painful than the immediate shock. In keeping with Quay's theory, psychopathic criminals preferred the immediate shock 5.5% of the time, while the other groups chose that alternative 87.5% of the time. Hare concluded that his psychopathic subjects desired the increased stimulation, even though it was seemingly more aversive.

Holding these neuropsychological deficits in mind, we will now examine additional bodies of literature that serve to further illuminate the connections between perceptions of contingency and causality (reward and punishment), justice and moral development.

Causality, justice and moral development

Experimental psychopathologists try to create in the laboratory, often using animals, conditions that mimic the onset and development of psychological disorders in humans. One area of research is the impact of exposure to uncontrollable experiences on human development. To expose an organism to an uncontrollable experience is to render it helpless; and to be helpless is to be in a condition wherein our actions do not influence what happens to us. In such circumstances the outcomes that we experience are under the control of arbitrary or random forces. Over the past three decades a great deal of research has been done on the impact of helplessness on individuals and groups.

In a typical helplessness experiment, the 'triadic design' is employed. This design enables researchers to expose one group of subjects to unpleasant controllable events, a second group of subjects to unpleasant uncontrollable events, and a third group to neither uncontrollable nor controllable events. What is illuminating about this design is that the subjects who are in the first two conditions (the controllable and uncontrollable conditions) are exposed to exactly the same amount of the aversive experience (for example, a loud buzzing noise) for exactly the same amount of time. When the subjects in the controllable condition figure out what they can do to turn off the noise, the noise goes off for the subjects in the uncontrollable condition as well. We say that the subjects in this latter condition are *helpless* because there is nothing that they can do to stop the noise. Their destiny, with respect to the noise, is determined wholly by the actions of another.

In the early stages of a typical helplessness experiment the subjects in all conditions will do all that they can to figure out how to avoid or stop the unpleasant stimulus. Sometimes they must solve a puzzle, or run through a maze, or jump over a barrier in order to turn it off or avoid it. In the uncontrollable condition subjects are exposed to situations in which it is impossible for them to solve the puzzle, make it through the maze or get over a barrier, but they do not know that the experiment is designed for them to fail. When subjects in this condition realise that their actions do not have an effect, they stop acting and begin to suffer the unpleasant stimulus passively. We have seen helplessness deficits develop in a wide range of species – including rats, cats, goldfish, cockroaches and humans – and thus we believe that the function of control is fundamental to life at all levels of existence.

The function of control is vital to so many species because it is connected with the fundamental law of cause and effect. As we suggested earlier, the operation of the law of causality is a manifestation of the principle of justice in nature. Because of the operation of this law, the natural world is rendered meaningful; that is, it is a place wherein the relationship between actions and outcomes can be discerned. Such discernment is a sine qua non of moral development. For organisms that have the cognitive capacity to prefer that some effects be realised, while others are avoided, causal effects take on hedonic value and may be experienced as rewards and punishments. While the expectation of reward and the fear of punishment are critical in facilitating all forms of human development, they play a particularly important role in moral development.

The importance of reward and punishment and hope and fear to moral development can be seen more broadly in the research of two social scientists who have developed the concept of *possible selves*. Hazel Markus and Paula Nurius have argued that people's willingness to delay immediate gratification and to work for important future goals is dependent upon assessments that they make about their possible future selves.[15]

Under healthy conditions everyone, according to the researchers, has a set of 'feared selves' and 'hoped-for selves'. A feared self might include the image of 'me in prison', while a hoped-for self might include the image of 'me as a competent father' or 'me as a doctor'. Importantly, this work has suggested that people must have *both* hopes and fears if they are to achieve important moral and professional goals. Young people who have feared selves ('me in prison') without corresponding hoped-for selves ('me as a good father') will generally not be deterred from crime and immoral acts by threats of punishment. The research by Markus and Nurius indicates that fear influences an individual's behaviour only if it threatens the loss of a valued possible self. Thus, if people can see no real options for becoming what they dream of, or if they have no vision of moral excellence that is sufficiently inspiring, they cannot be prevented from committing crimes or immoral acts by increasing severity of threats against them.

We believe that punishing individuals in such a condition often generates a paradoxical effect: humans who are chronically controlled by fear without an attending sense of hope tend to become outwardly passive. But beneath this thin veneer of passivity there often boils a sense of anger, hostility and rage that finds expression in brutal acts of violence directed against innocents. Investigations into the recent incidents in the USA involving youths who murdered several of their teachers and classmates revealed that the murder sprees were preceded by long periods of bullying, ostracism, parental mistreatment and other forms of abuse that did not receive proper consideration from those who possessed the power to stop it.

❋ ❋ ❋

People whose hoped-for selves cannot be realised grow to disregard the justice-related principles that govern community life because they do not expect to derive the benefits associated with respecting the rights of others. And while it is true that 'virtue is its own reward', the recognition of this truth requires a degree of moral development generally made impossible by chronic exposure to injustice. Furthermore, in the absence of viable options for exercising freedom, the threatened loss of freedom is meaningless. Within many inner-city and poor rural communities the loss of hope has resulted in the eclipse of fear; and without hope and fear it may be impossible to either constrain destructive impulses or unleash potential. The consequence is often lawlessness and the creation of a crucible which encourages the worst, most sociopathic elements of society.

Michael Tonry, the author of *Malign Neglect: Race Crime and Punishment in America*,[16] reports that in the USA the number of African-Americans in prison has tripled since 1980; between 1979 and 1992 the percentage of black people among those admitted to state and federal prisons grew from 39% to 54%; incarceration rates for black people in 1991 were nearly seven times higher than those for white people; and in 1991, in the nation's capital, 42% of black men aged 18–35 were in jail, on parole or awaiting trial. The figure for Baltimore was 56%. To some, these high rates of incarceration suggest the presence of bad genes or irremediable character flaws; to others, they suggest a need to look more carefully at the justice of present day society. (While distinctions must be clearly made between antisocial personality disorder and criminal behaviour, it is clear that prisons can be powerful incubators for exacerbation of the disorder.)

We cannot make sense of antisocial behaviour without appreciating the human need for a sense of direction in life. Without a set of goals to strive for which transcend the self, and which equip an individual with a sense of meaning and purpose, the individual tends to become bored, depressed and angry. These feelings, in turn, can fuel processes of interpersonal violence and destruction that are more savage in the human world than in anywhere else in nature.

From our perspective, then, any discussion of complex problems such as antisocial personality must take into consideration biological processes, personal histories, as well as the broader social, philosophical and spiritual dimensions of human life.

References

1 Lykken DT. *The antisocial personalities*. Mahwah, NJ: Lawrence Erlbaum Associates, 1995.
2 Black D. *Bad boys, bad men*. New York: Oxford University Press, 1999.
3 Wetzell RF. *Inventing the criminal: a history of German criminology*. North Carolina: University of North Carolina Press, 2000.
4 Rafter NH. *Creating born criminals*. Chicago: University Press, 1997.
5 Prichard, JC. *A treatise on insanity and other disorders affecting the mind*. London: Sherwood, Gilbert and Piper, 1835.
6 Doren DM. *Understanding and treating the psychopath*. New York: John Wiley, 1987.
7 Patridge, GE. Current conceptions of psychopathic personality. *Am J Psychiatry* 1930:10;53–99.
8 Cleckley H. *The mask of sanity*. Saint Louis: CV Mosby Company, 1941.
9 Hare RD. Performance of psychopaths on cognitive tasks related to frontal lobe function. *J Abnorm Psychol* 1984:93;133–40.
10 Hare RD. Temporal gradient of fear arousal in psychopaths. *J Abnorm Soc Psychol* 1966:70;442–5.
11 Hare RD, Shalling D. *Psychopathic behaviour: approaches to research*. Chichester: John Wiley, 1978.
12 Lykken, DT. A study of anxiety in the sociopathic personality. *J Abnorm Soc Psychol* 1957:55;6–10.
13 Quay HC. Personality and delinquency. In H Quay (ed) *Juvenile delinquency*. New York: Litton, 1965:pp1–17.
14 Fairweather GW. *Serial rote learning by psychopathic, neurotic and normal criminals under three incentive conditions*. Chicago: University of Illinois, unpublished doctoral dissertation, 1953.
15 Markus H and Nurius P. Possible selves. *Am Psychol* 1986:41;954–69.
16 Tonry M. *Malign neglect: race, crime and punishment in America*. New York: Oxford University Press, 1995.

9

Universal values

WILLIAM HATCHER[†]

Professor Hatcher argues that there are human values and traits that are universal and therefore not due to socialisation. He acknowledges that acquired knowledge (socialisation) profoundly affects value judgements but it does not follow that all judgements are wholly based on acculturation. There is a primal, transcultural aspect to humanness which constitutes a metaphysical truth. Part of this humanness is an intrinsic essential moral sense.

Everyone acknowledges that humans have preferences — that we make judgements of worth or value about our experience of life. Whatever we may actually say or think about our value judgements, it is our actions which reflect them most faithfully. I may say I do not like chocolate, but if I regularly eat large quantities of it, without any external duress, you would be most reasonable to conclude that I do in fact like chocolate.

More generally, the sign of a positive value judgement is our attempt to repeat the valued experience, consistent avoidance behaviour being the corresponding sign of a negative judgement. It is sometimes held that there is no ultimate, common basis for value judgements — that they are arbitrary. According to this view, an individual may, depending on his life circumstances and his reaction to them, come to prefer (or value positively) anything: pain, cruelty, suffering, death, the ugly or the hideous. Those who support this view usually do so by citing cases of individuals who do indeed seem to have exhibited such preferences, eg sadomasochists.

But there is a fundamental flaw in this kind of argument. There is of course no doubt that individual differences in preference – even extreme individual differences in preference – may be shown to exist in certain matters. But such differences do not in themselves refute the idea that there may be a more

---

[†]Sadly, William Hatcher died before this book was published.

fundamental, underlying, universal basis for most preferences, and that deviations from them take place only within certain limits and under extreme circumstances.

Similarities and differences

We live in a world that is rich with difference and multiplicity. Any two entities in existence, or 'existents', may be compared according to their similarities or their differences.

Moreover, there is a question not only of the objective similarities and differences between two existents but also of our subjective perceptions. There is the question of the relative importance (value) we assign to differences versus similarities in any given case. This is the question we will discuss in this chapter.

A multiplicity of examples show conclusively that cultural (socially learned) values are not the ultimate values. Indeed, socialisation and individual learning build upon fundamental, innate value preferences.

Those who insist that cultural relativism is primary, might point out that it is possible to train and condition someone to the extent that they prefer quinine to honey, or even find honey loathsome and repugnant. Such examples show that natural value preferences can be altered by socialisation, but they do not constitute an argument against the existence of primal, natural value preferences in the first place. Or, to put it another way, such a socialisation would be universally recognised as an exception, one which would require explanation (how is it that this person has acquired such a distaste for honey?). But if told that an individual loves honey, we would not require an explanation, because that is the 'natural' state of affairs (the 'default position' as it were).

In the example of honey and quinine, the reaction of an infant would be spontaneous and instinctive. Nonetheless, it can be legitimately separated into at least the following three stages:

1 There is an objective difference between honey and quinine. This is due ultimately to objective differences in their molecular structure.
2 The human sensori-neural apparatus has the capacity to detect or experience this objective difference in some manner.
3 Relative value is given to the difference – the fact that the encounter with the substances is experienced positively (and to a certain degree of intensity) or else negatively (to a certain degree of intensity).

The transition from objective to subjective takes place in the second stage. The objective difference between the two substances is translated into or reflected by a difference in subjective, inner states provoked by encounters with the two substances. However, this subjective difference in experience does not in itself imply a discriminating value judgement. The two stimuli both induce reactions of either positive or negative valency, or memory to avoid further encounters. There would be either no value discrimination, or else a value discrimination in the relative intensity of the motivation to approach or to avoid the stimulus.

In any case, the point is that the third of the above stages, in which a value judgement is made, is distinct from the second, in which there is a subjective difference in the quality of the experiences induced or provoked by the two encounters. In other words, the making of value judgements presumes, or is based upon, the ability of the organism to discriminate between two different experiences, but the value judgement itself cannot be reduced to the simple fact of difference.

We perceive our experience as positive (relatively pleasant) or negative (relatively unpleasant). In other words, value judgements are not inherent in (primal) experience itself but arise from the consequences of experience. In the simple instance above, the consequences are simply the emotions 'pleasant' on the one hand and 'unpleasant' on the other.

It is obvious that more complex experiences will give rise to more complex consequences. In particular, once the individual has matured to the point of acquiring not only sensori-neural sensibility but also human self-awareness, the third, evaluative stage in the list above can become more explicit and thus more autonomous. For example, the mature and autonomous human being may be able to say, of an initial experience of drug euphoria, 'This was intensely pleasant but dangerous for my ultimate well-being and should be deliberately avoided in the future.'

Value judgements of this sort are very sophisticated and definitely involve a significant degree of conscious knowledge, both of the self and of reality. But they still fit the basic paradigm above, namely that the individual's value judgements are based on the consequences of experience. It is just that the self-aware subject has a certain knowledge not only of the short-term consequences of the experience (eg that it feels pleasant) but also of the longer-term consequences (eg that repeating the experience can lead to drug dependency and thus a significant loss of autonomy).

Thus the fundamental point made a few paragraphs above is repeated and amplified. The fact that knowledge (socialisation) profoundly affects value judgements does not mean that all value judgements are arbitrarily or wholly social in nature, because many value judgements are rooted in that primal experience which means that we all naturally perceive various aspects of reality as relatively pleasant or relatively unpleasant. This primary, binary experience of pleasant/unpleasant (pleasure/pain, good/bad) is rooted in essential and universal human nature, and is thus transcultural. In other words, there may indeed be value judgements that are arbitrarily generated by a given culture (by the process of socialisation) but there are also value judgements that are universal and transcultural.

The examples we have given so far might be said to involve only the most primal physical instincts of man. One could still ask whether the higher-order or moral value judgements are not wholly cultural in nature and have no primary base. Given that there are universal value judgements, are there universal moral values?

Moral values

When we speak of 'moral values' we are still speaking, first of all, of values generated by the three-stage process mentioned above. Specifically 'moral' values, however, are those values which arise primarily from interhuman (or 'social') interactions and which involve judgements about how we experience both ourselves and others. The universality of such judgements arises from the fact that all humans experience love and kindness positively and experience cruelty and hatred negatively. This universality is, again, rooted in essential human nature.

That humans respond positively to love, acceptance and kindness is not just a dictum of moralists but a fact of human nature. For example, Sigmund Freud was an atheist who held an extremely negative view of human nature and intrinsic human potential. He was anything but a moralist. Yet all of his observations and theories support the thesis that the human personality is significantly determined by early experience, and particularly by the quality of early interhuman relationships, beginning with the mother and moving out gradually to the father and other significant adults. The thrust of his findings was that children who receive love, acceptance and nurturing from these significant adults are relatively healthy and happy, and those who are subject to the trauma of rejection, abuse, hatred or aggressive cruelty generally suffer its negative effects for the rest of their lives.

These initial findings of Freud have been validated and revalidated by a host of other psychologists, using many different approaches. But does anyone really doubt that intrinsic human nature responds positively to love and kindness and negatively to hatred and cruelty? Who has not experienced the warmth of being loved in contrast to the anxious knot in the stomach when the object of aggression, insult or rejection?

Moreover, this essential human nature is the ultimate source of all value judgements, whether positive or negative. This observation suggests that the highest value in creation (the highest value in existence other than God) is that intrinsic and essential human nature from which all value judgements flow. That there is such a universal human nature is a Platonic hypothesis that can be confirmed but not proved by observation alone. Let us examine this more carefully.

The Platonic underpinnings of universal morality

We observe that there are certain stimuli to which all but a negligible minority of infants respond positively – honey and love, for example – and other stimuli to which all but a negligible minority of infants respond negatively – for example, quinine and cruelty. Does this not prove that there is indeed a universal human nature?

It certainly suggests strongly that there *may* be a universal human nature, but to really answer the question in the affirmative, we have to consider the metaphysical basis of human nature itself. If we are materialists who hold that humans are just a particularly evolved species of animal – whose nature at any moment of evolution is totally determined by the physical parameters then in existence – then the answer to our question may well be 'no', because under such an hypothesis we cannot exclude the possibility that the physical parameters will change in such a way that these intrinsic value responses are significantly altered. Mutations could be so drastic that there would be no uniformities whatsoever in our spontaneous value responses to stimuli.

All available evidence suggests that there has been no real change in human nature over the past, say, ten thousand years, so such a drastic change does not seem very likely in the foreseeable future. But that is not the point. The point is that if we attribute present uniformities in spontaneous human value responses solely to a fortuitous genetic configuration, then we cannot consistently talk about an essential human nature. We will have only human nature today, human nature tomorrow, etc. We cannot make any general statements about what is essentially human, since we have no guarantee that some subtle, even apparently trivial, genetic mutation could alter something we now consider essential.*

***Editor's note** Craig and Loat's conclusions partly support this in that there are specific genetic variations which affect the behaviour of individuals such as schizophrenia. However, they cite the DRD4 gene as significantly changing human behaviour, in a broad contingent over 40,000 years or so.

Since such alterations have not happened, we therefore posit, as a fundamental metaphysical truth, that there does exist an intrinsic, essential, universal human nature, and that observed uniformities and regularities in spontaneous human value responses to external stimuli reflect, albeit imperfectly and approximately, this human nature. In Platonic terms we are positing the objective existence of humanness. We can therefore in principle accept the extrapolation of scientific findings relating biology to behaviour.

The fundamental (but not exhaustive) characteristics of human nature are:

- consciousness (the existence of a subjective world of conscious inner states within each individual)
- mind (the capacity of this conscious subjectivity to reflect or model, if not perfectly then at least significantly, the structure of the world outside subjectivity)
- heart or affectivity (the capacity to feel certain emotions or subjective sensations, most particularly the capacity to experience the emotion of altruistic love*)
- will and intentionality (the capacity to contemplate and execute certain courses of action).

We might speak of cognitive consciousness, affective consciousness and volitional consciousness.

All human values and value preferences can be consistently regarded as generated by a suitably combined interaction of the fundamental human capacities of consciousness, mind, heart and will.

The force of love

The human response to the recognition of value is a complex of thoughts, feelings and actions that we call love. Our 'hearts' feel deep emotions of attraction towards the valued entity. We want to move closer to it, to possess it if possible or else to establish a harmonious relationship with it. We want to know everything we can about the object of our love. We are fascinated with every facet of it. And we are moved to act so as to enhance and/or serve the valued object.

Love, then, is a force which inhabits us to the degree that we appreciate true value. Love is the response of the human being to the perception of value. If the perception of value is an illusion, then the love will ultimately prove false. But if the perception is true, then love will grow and develop. In

---

***Editor's note** This is supported by the work of various neuroscientists, who, for example, indicate the hardwired elements of neurocircuitry responsible for empathy.

particular, true or authentic human relationships are based on a true perception of the intrinsic value each of the other (and of the self). What we truly perceive is nothing less than the essential, intrinsic and universal humanity which each of us possesses. In particular, I know that what makes me suffer will probably make you suffer, and what makes me happy will probably make you happy. I will therefore feel compassion for your suffering and gratitude for your genuine happiness. Moreover, I will shrink from being the deliberate cause of suffering on your part.

Relationships based on genuine love are necessarily symmetrical, because they are based on the mutual recognition of an intrinsic and universal value, a value shared by both parties. Once I achieve the ability to recognise universal humanness, I can recognise its manifestations everywhere, both in myself and in others.

We may ask, in turn, what is the greatest indicator of genuine love in a relationship, and the answer is simply how we actually treat others.

Hierarchies of value

The value-supremacy of essential human nature is only the highest step in a continuum of objective values that are each inherent in the structure of reality, from minerals to plants to animals.

Humans, more than minerals, plants and animals, can also process energy in its most refined form – that is, as abstract (symbolic) information. Humans have the ability to attribute to arbitrarily chosen symbols a meaning or significance totally unrelated to the physical form or structure of the symbol itself.

This cumulative multifunctionality of higher organisms gives us a more precise understanding of why the human being stands at the apex of the value hierarchy. Such an understanding has become especially important since the advent of computers that can outperform human mental functioning in certain specific respects. Why, as some philosophers of artificial intelligence have argued, could some future, sophisticated computer not be of equal or greater value than humans?

The answer is two-fold. On the metaphysical level it will forever be the case that it is humans who have created computers, not the reverse. Electronic computers, however sophisticated, were not a naturally occurring phenomenon but had to be conceived abstractly by the human brain before they existed concretely. If we accept the highly plausible philosophical principle that a cause must always be greater than its effect (which, in fact, is just a metaphysical form of the law of entropy), then humans will always be of greater value than computers or any other creation of the human mind.

Faith in the superior value of human beings has, for some people, been shaken by such things as the recent defeat of Gary Kasparov by the computer program Deep Blue. As for myself, I am quite willing to accept that it is possible to program a sufficiently complex computer to outperform human functioning in any given specific area of endeavour. Indeed, this is already the case for human physical performance: our artificially created machines are faster, quicker, stronger than any humans. The superiority of human functioning lies precisely in its seemingly inexhaustible multifunctionality. A single human organism has genuine self-awareness, can love, play the violin, do mathematics, invent computers, play tennis, reproduce, etc. The very definition of the human is that he or she is forever undefinable.

Clearly we will never be able to invent a robot that can, alone, accomplish all the various tasks that are accessible to the ordinary human being. For, if we could create such a machine, it would, by definition, be the human being: we would have recreated ourselves, a highly implausible if not logically impossible achievement (again in view of the known laws of systems dynamics and, in particular, the law of entropy).

Thus, in the final analysis, morality and moral values arise on the one hand, from the existence of an objective value hierarchy that is embedded in the very structure of reality and on the other hand, from the universality of essential human nature, which allows us to apprehend this value hierarchy and act upon this understanding, if of course we choose to do so.

10

The meaning of the 21st century

JAMES MARTIN

Dr James Martin demonstrates the need for a change in human attitudes and behaviour. He deploys a powerful argument supporting fears for the Earth's survival and relates this to the immorality of short-termism, globalism and the acquisitive society. The identification of macrotrends and data from demographic studies only now available because of computer power and new technology enables accurate predictions and assessment of lead times. These are largely pessimistic but there are 'leverage factors' which, if used adroitly, may balance the negative effects, at least to some extent, of greed, lack of education, vested interests, excessive consumerism and bad governance.

When evolution stumbles into something fundamentally new, like the first flower, it usually does not get it right the first time. It produces mutations on the theme. The first flowering plant species probably did not last long. The Cambrian period produced numerous fascinating creatures that did not survive. An alien professor from a planet far, far away might observe us on Earth suddenly going wild with technology, and ask, 'Can this survive?' He or she (or it) might say to students, 'The people on Earth clearly haven't got it right yet. But they are intelligent. Will they use their intelligence to make corrections? Will they make the corrections in time, or will they self-destruct? If they get it right, will there be the equivalent of flowers everywhere? Or will this be one of evolution's unsuccessful experiments?'

Fifty years from now we may know the answers to the professor's questions. So the lifetime of people who are teenagers now will be a fascinating era.

Macrotrends indicating future problems

Key aspects of the future can be sketched by identifying macrotrends – the main patterns of change that are foreseeable. A macrotrend is an ongoing trend that has substantial consequences and that seems inevitable or almost inevitable. For one example, the growth of the population of the planet can be estimated. For another, environmentalists predict the growth of carbon dioxide emissions, under different circumstances. At present, we are pumping more carbon dioxide into the atmosphere than the Earth's forests can absorb, and every year we destroy 44 million acres of forest. Not all macrotrends are negative; some are beneficial and we will take advantage of them, such as a growing use of new energy sources and rapidly increasing bandwidth of telecommunications.

One can identify many macrotrends and together they form the skeleton of the future, along with demographics and technology that is predictable because of the long lead-time from research to reality. We can put flesh on to the skeleton in a variety of different ways. Some macrotrends are obvious, such as the decline of ocean fisheries. Some are clear to experts in the field, such as Gordon Moore's famous prediction, 'Moore's Law',[1] about the long-term increase in the number of transistors on a silicon chip. This needed a deep understanding of the chip-manufacturing technology. Much can be predicted about the future impact of technology – for example, the globalisation of new media and some of the consequences of genome mapping.

When the effects of macrotrends are combined, it is clear that we are in deep trouble. Let us look at some examples. Every six weeks we add people equivalent to the population of New York to the planet and exterminate about 4,000 species. World population will increase by about three billion in the next three or four decades, but every year we lose 100 million acres of farmland, lose 24 billion tons of topsoil, create 15 million acres of new desert, deplete the stock of fish in the oceans – nearly 90% of the most prized fish in the oceans are gone – and fishing fleets pump 22 million tons of dead 'bycatch' back into the sea. Water is essential for growing food. Much of the water humankind uses comes from large underground aquifers and dates back many ice ages. When the ancient water is used up, we will have to live on rainwater. Today humankind is using about 160 billion tons more water each year than is being replenished by rain. This water deficit will cripple industry and decimate populations.

There are accelerating differences between rich and poor nations. The wealthiest flaunt their materialism in the face of the poorest via media and the forces of globalism; the sudden opening of an age of terrorism reflects the anger of the have-nots. We cannot have a world with nine billion people wanting a lifestyle like those they watch on television because such consumption, if not redirected, will shoot far beyond the carrying capacity of the planet. The question arises: which of these macrotrends will be the critical terminal factor?

Possible solutions

The good news is that we are steadily realising that there are solutions to most of these problems. The term 'leverage factor' is used to refer to relatively small and politically achievable actions that can have powerful results. For example, in poor countries, when women learn to read they have fewer children, and it is relatively inexpensive to achieve a higher literacy rate. There are many situations where a relatively small action, or minor change in rules, can cause massive changes in outcome. In every aspect of the planet's problems, there are powerful leverage factors, many of them far from obvious. In aggregate they give us the capability to slowly transform the planet.

The forces which characterise the near future are so large that they will inevitably change Earth's different civilisations. The 21st century will be characterised by an extraordinary growth in knowledge and techniques for putting knowledge to work. An incomparably higher quality of life than today could be achieved without causing planetary harm if we make the right choices. Once we address and solve the question of how we evolve civilisation, there will be much greater flexibility for dealing with the planet's issues. The subject then changes from grappling with intractable problems to inventing a set of options for a worthwhile future.

We should be asking: what are the forms of civilisation that we could create with our new wealth and new technology? As in the days of Thomas Jefferson, a great debate should be fostered about future civilisation. We need to ask what future civilisation could be like. What principles should guide our changing civilisation? The decades ahead could be the most interesting eras in humankind's history because of what is now possible.

Given our advancing knowledge, if we fail to create a great civilisation it will be because of avoidable problems caused by greed, lack of education, extreme vested interests and bad governance. It may be that companies drive excessive consumerism and maximise profits by focusing on the lowest common denominator. It may be because false mythologies prevail instead of science. It may be because the West is stuck in its past. To focus on planning our future civilisation is our obligation to our progeny and the planet.

The Astronomer Royal gives 'fifty–fifty odds' that humankind will exterminate itself by the end of the century.[2] We are the first people in history to have had the capability of saving or rapidly eliminating humanity. Weapons will become ever more devastating, and the Earth's population will become desperately overcrowded and short of critical resources. At the same time, civilisations and religions with radically opposed but passionately held views are being brought face to face by the media, travel and global commerce. Export of radically different ideologies has provoked an era of terrorism. Unfortunately it coincides with weapons of mass destruction becoming easier and cheaper to produce.

Now in the 21st century, the planet is unstable because of the possibility of population and development overshooting the limits of the planet, and technology is barely controllable. Because technology will become powerful enough to destroy us, we need to learn to steer and constrain technology.

There has to be a transition from the present untenable course to a world where we control these potentially devastating forces. I shall refer to this as the 'survivability transition'. A critical part of this transition is the move to sustainability – where we are no longer using up resources that cannot be replaced. We need to achieve a sustainable civilisation in which the highest quality of life is achieved without planetary destruction. However, sustainability alone is not enough. Achieving sustainable development gives no guarantee that we will survive.

Many changes are needed to put humankind on a path to survivability. There is, however, not much time. The main ocean fish stocks will slowly recover, as some whales did, if we take action now, but if we continue the current rate of destruction, seas, one after another, will become too depleted to recover. Sooner or later, agriculture will experience three bad harvests in succession in developing countries. If at that time we have not put aside the large reserves of food needed for food security, there will be famine on a far greater scale than ever known. In many of these areas there is a race against time.

The task of the 21st century is to achieve survivability – to eliminate the factors that could destroy humanity. Many actions are needed and the many powerful leverage factors available to us must be thoughtfully employed. Time is limited, and the longer we delay, the more difficult the problems and the greater the risks.

Today's young people are a transition generation – hopefully making the transition from fatal consumption to survivability – but our education system is not preparing the transition generation for the role they need to play.

The technology that allowed us to damage the ozone layer, build atomic bombs and start the polar ice caps melting was brute-force technology – the internal combustion engine, coal burning, crude chemical pesticides and smokestack industry in general. The transition era will have clean new technologies such as hydrogen cars, genetic engineering, nanotechnology, chain reactions of computer intelligence and biological remedies. If we use these new capabilities with wisdom, the planet will survive.

I believe that the problems of water extraction, pollution, global warming, overpopulation and international conflicts can be solved and that the decades of my daughter's lifetime will be a most exciting time to be alive. To ensure this, humankind must understand why we have these problems. The subject needs to be clarified and the many leverage factors thoroughly researched. If the James Martin Institute for Science and Civilisation grows as intended, it will become a resource of immense value to humankind in the quest for solutions.*

The alien professor from the planet far, far away will say, 'Now here's a situation worth watching. The people of Earth have reached a point where they could destroy themselves. Amazingly, they have numerous business schools, but no school for survival.'

In pursuit of moral excellence

The incidence of alcohol and drug abuse, violence and crime coupled with poverty, prejudice, corruption, injustice and HIV/AIDS has laid a great burden on humanity. The governments of the world are searching for solutions. The main focus has been on law enforcement, security, prison expansion and information campaigns, but despite the human and financial resources expended, these problems continue with increasing intensity.

Social ills are preventable through character education, which is most effective in the formative years. It is during this time that moral values are internalised and character is formed. Indeed, there are two forces which govern human behaviour, the internal force (moral) and the external force (law, police, prisons, etc). It is the moral force that enables an individual to make moral decisions and act upon them, whereas the external force is designed to protect and punish and not to improve the character of the individual.

In the past five years the author has met with the heads of state, royalty (including His Royal Highness the Duke of Edinburgh), ministers of health, education and justice (including three ministers of education in the UK), Members of Parliament, leaders of thought and religious leaders in 55 countries, recommending the inclusion of moral education in the school curriculum. Moral values such as truthfulness, trustworthiness, kindliness, respect, unity, justice, fairness, compassion, generosity and the pursuit of excellence constitute the building blocks of the character. They are universal values, which are acceptable to all cultures, traditions, religions and nations. The

---

*The James Martin Institute for Science and Civilization at Oxford University was launched in 2004, and is one of 11 research entities of the James Martin 21st Century School. All are committed to an interdisciplinary approach towards finding solutions to the biggest problems facing humanity and identifying the key opportunities of the 21st century (see www.martininstitute.ox.ac.uk/jmi and www.21school.ox.ac.uk).

response of the government representatives was overwhelmingly positive. They agreed that moral education is an indispensable part of education beginning from earliest childhood and that developing moral capabilities is about educating children so that they think about the welfare of the community, participate in collective enterprise, manage their affairs with responsibility and good conduct, build unity, seek and appreciate diversity, take initiative, persevere to overcome obstacles and search for solutions. It awakens their consciousness about their own nobility of character and their capacity for personal/group decision-making and self-reliance.

In the past the underlying cause of social ills was considered to be economic. However, the World Bank has realised the futility of achieving economic development without a moral and spiritual foundation. Hence they have created a Department of 'Interfaith Dialogue' to explore the integration of moral and spiritual principles into the development paradigm. Lack of truthfulness and trustworthiness in human affairs, for example, has had catastrophic results, evidenced by the fall of Enron and World.Com with losses of billions of dollars.

The United Nations (UN), in its 50th Anniversary Declaration, stated that one third of humanity is deprived of four basic human rights: education, food, healthcare and clean water. The UN estimated that $40 billion dollars a year would be needed to address this deprivation. In a commentary in the *New England Journal of Medicine*,[4] it was stated that $40 billion dollars is less than 4% of the assets of the 225 richest people of the world. It concluded that the crisis is not economic but moral and that voluntary sharing was the proposed solution for the poor distribution of wealth.

The High Commissioners of Human Rights have agreed that teaching and promoting the oneness of humankind with its emphasis on acceptance and respect for all nations, religions, races and people will largely eliminate human rights abuses. The timeliness of moral education cannot be overestimated.

References

1. Moore G. Cramming more components onto integrated circuits. *Electronics*. April, 1965.
2. Rees M. *Our final hour*. New York: Basic Books, 2003:228.
3. United Nations. *Fiftieth Anniversary of the Declaration of Human Rights*. See www.un.org/rights/50/anniversary.htm
4. *New York Times*. Kofi Annan's astonishing facts! September 27, 1998:16. Cited in Human Rights and Health – the Universal Declaration of Human Rights at 50, *NEJM*, 1998:339;1778–81.